An International Directory of Building Research Organizations

Building Research Board
Commission on Engineering and Technical Systems
National Research Council

NATIONAL ACADEMY PRESS
Washington, D.C.   1989

National Academy Press • 2101 Constitution Avenue, N.W. • Washington, D.C. 20418

NOTICE: The project for which this directory was prepared was approved by the Governing Board of the National Research Council, whose members are drawn from the councils of the National Academy of Sciences, the National Academy of Engineering, and the Institute of Medicine. The members of the committee responsible for the project were chosen for their special competences and with regard for appropriate balance.

This directory has been reviewed by a group other than the authors according to procedures approved by a Report Review Committee consisting of members of the National Academy of Sciences, the National Academy of Engineering, and the Institute of Medicine.

The National Academy of Sciences is a private, nonprofit, self-perpetuating society of distinguished scholars engaged in scientific and engineering research, dedicated to the furtherance of science and technology and to their use for the general welfare. Upon the authority of the charter granted to it by the Congress in 1863, the Academy has a mandate that requires it to advise the federal government on scientific and technical matters. Dr. Frank Press is president of the National Academy of Sciences.

The National Academy of Engineering was established in 1964, under the charter of the National Academy of Sciences, as a parallel organization of outstanding engineers. It is autonomous in its administration and in the selection of its members, sharing with the National Academy of Sciences the responsibility for advising the federal government. The National Academy of Engineering also sponsors engineering programs aimed at meeting national needs, encourages education and research, and recognizes the superior achievements of engineers. Dr. Robert M. White is president of the National Academy of Engineering.

The Institute of Medicine was established in 1970 by the National Academy of Sciences to secure the services of eminent members of appropriate professions in the examination of policy matters pertaining to the health of the public. The Institute acts under the responsibility given to the National Academy of Sciences by its congressional charter to be an adviser to the federal government and, upon its own initiative, to identify issues of medical care, research, and education. Dr. Samuel O. Thier is president of the Institute of Medicine.

The National Research Council was organized by the National Academy of Sciences in 1916 to associate the broad community of science and technology with the Academy's purposes of furthering knowledge and advising the federal government. Functioning in accordance with general policies determined by the Academy, the Council has become the principal operating agency of both the National Academy of Sciences and the National Academy of Engineering in providing services to the government, the public, and the scientific and engineering communities. The Council is administered jointly by both Academies and the Institute of Medicine. Dr. Frank Press and Dr. Robert M. White are chairman and vice-chairman, respectively, of the National Research Council.

Work on this directory was initiated under Contract No. 1030-562112 between the National Academy of Sciences and the State Department. Work was completed as part of the technical program of the Federal Construction Council (FCC). The FCC is a continuing activity of the Building Research Board, which is a unit of the Commission on Engineering and Technical Systems of the National Research Council. The purpose of the FCC is to promote cooperation among federal construction agencies and between such agencies and other elements of the building community in addressing technical issues of mutual concern. The FCC program is supported by 14 federal agencies: the Department of the Air Force, the Department of the Army, the Department of Commerce, the Department of Energy, the Department of the Navy, the Department of State, the General Services Administration, the National Aeronautics and Space Administration, the National Endowment for the Arts, the National Science Foundation, the U.S. Postal Service, the U.S. Public Health Service, the Smithsonian Institution and the Veterans Administration.

Funding for the FCC program was provided through the following agreements between the indicated federal agency and the National Academy of Sciences: Department of State Contract No. 1030-621218; National Endowment for the Arts Grant No. 42-4253-0091; National Science Foundation Grant No. MSM-8600676, under master agreement 82-05615; and U.S. Postal Service grant, unnumbered.

Library of Congress Cataloging-in-Publication Data

An International directory of building research organizations.
    Bibliography: p.
    Includes index.
    1. Building research—United States—Directories. 2. Building research—Directories.
I. National Research Council (U.S.). Building Research Board.
TH23.I58   1989   690'.72   89-12320
ISBN 0-309-04027-2

Copyright © 1989 by the National Academy of Sciences

No part of this book may be reproduced by any mechanical, photographic, or electronic process, or in the form of a phonographic recording, nor may it be stored in a retrieval system, transmitted, or otherwise copied for public or private use without written permission from the publisher, except for the purposes of official use by the U.S. government.

Printed in the United States of America

Foreword

In this often fractious world, all nations share the needs for buildings and facilities of physical infrastructure to shelter and productively support the activities of their people. We all strive to improve the quality of our lives, and so these needs persist, regardless of levels of national income, political philosophy, and social structure.

Building research, the directed effort to discover new and better ways to meet these needs for buildings and physical infrastructure, thus offers benefits to all and occurs in many nations around the world. Opportunities for improving the materials, processes, and products of building far exceed the resources available for research, and so researchers and policy makers must strive to focus their attention on those matters of greatest concern and where their efforts are most likely to have valuable results. Knowledge of who the researchers are and what they are doing will help the community of researchers to achieve this focus. This knowledge will in turn help the beneficiaries of building research—all of us—by encouraging more rapid improvement of the built environment for all people.

The Building Research Board (BRB) has produced *An International Directory of Building Research Organizations* to help spread this knowledge. While our context is global, we must acknowledge a parochial interest as well: Building research and technological innovation in the United States are lagging. They are lagging in comparison with the great strides being made in such fields as electronics and bio-technology that may have application in building. They are lagging in terms of research spending, compared with the importance of construction in the U.S. economy and compared with other nations. Our lagging research effort threatens the productivity and competitiveness of our construction industry in an increasingly global marketplace.

We need, as a nation, to be aware of what our partners and competitors are doing, so that we may take advantage of the work of others that can help us to solve our own problems, bring our achievements to the attention of others who may find them useful, and maintain our economic leadership. The BRB hopes that this directory will in its small way help us to achieve these ends.

Andrew C. Lemer, *Director*
Building Research Board

Preface

Given the fact that the U.S. building industry is the nation's largest industry with over $400 billion in revenues, the amount of research supporting it is small, concentrated in a limited number of locations and fields, and, in some cases, closely held in corporations. In addition, most people in the building industry have little knowledge about research going on in the United States, and even less about research being undertaken abroad.

Unlike research performed within the contexts of other large industries, building research is highly fragmented, reflecting the nature of the industry. Knowledge and theory tend not to be built on an industry-wide basis. Because of these characteristics, there is no industry-wide source of information on the knowledge base of the building industry in the United States and abroad. This directory narrows that gap by making available in one place a compilation of U.S. and international organizations involved in building-related research.

In its report, *Building for Tomorrow*, a committee of the Building Research Board called for greater recognition of the global aspects of the building industry of the future. This directory is a first step to help inform the building industry and others about the types of organizations and the diversity of work ongoing around the world.

Originally, this effort was part of a larger one. The U.S. Department of State, as part of a program undertaken with the Building Research Board to develop security-related criteria for future embassy buildings, requested a report on where technological advances in the building industry were taking place in order to assure itself of the highest level of building expertise. To do so, the agency needed a mechanism for informing itself about current building-related research, and for acquiring the research support its special needs require. The collection and organization of information about the nature, extent, and depth of the knowledge base for building research resulted in this publication.

Several studies have addressed the need for industry-wide information on research and a number of groups have supplied some of this information. Worth special note are the Architectural Research Centers Consortium's (ARCC) *An Agenda for Architectural Research, 1982*; a study by the Franklin Research Center for the National Institute of Building Sciences called *Existing Systems for the Identification, Determination, and Communication of the Research, Development, and Information Needs of the Building Industry*; and *Proceedings of the Building Industry Roundtable on Technology Transfer and Research Utilization*, prepared by the American Society of Heating, Refrigerating, and Air-Conditioning Engineers (ASHRAE) for the Department of Energy.

The organization profiles are organized into two major sections: United States and international. Within the U.S. section, the profiles are subdivided according to the type of organization performing the work: associations, corporations, universities (both architecture and engineering departments), federal laboratories, and other research centers.

The associations contacted either provided information about their own research, or made referrals to research facilities. Most of the associations contract out research. The corporations included here are only those that do, or are willing to undertake, outside contract research. For the most part, these corporations have established research and development (R&D) centers. Building research takes place in 46 universities and technological institutes. Multiple research centers within the same university are reported separately. Frequently, these centers have access to each other's equipment and facilities, and are associated in a formal or informal interdisciplinary research effort. The profiles of federal laboratories describe only the facility's building-related research. A final section comprises independent, non-profit research centers and laboratories.

The international section of building research organizations is arranged alphabetically by nation. Organizations from 53 nations responded to a survey on building research activities. The results of this survey form the basis of the international section.

Each profile—United States and international—contains the name and address of the organization and a contact person(s) within the organization. The mission, focus of research, and primary work are detailed for each organization. Publications, where appropriate, are listed. For U.S. organizations, distinctive attributes of the organization, such as research laboratory equipment or computer technology, are given. For international building research organizations, the sources of finances are given.

Finally, a key word index can be found on page 213 to direct the reader to organizations undertaking research in different areas.

Many individuals helped to gather information and put it in a format useful for this directory. J.F. Coates, Inc. managed the team that drew together the domestic section. In addition to Joseph Coates, members of the team, whose assistance is gratefully acknowledged, included Maria Gladziszewsky, Jennifer Jarratt, Darold Johnson, Bill Neufeld, and Lydia Perry. Karen Burdett, a research summer intern with the Building Research Board, developed and managed the international survey that resulted in the international section of this directory. In this effort, she was directed by Noel Raufaste, who was responsible for international efforts of the board. John Eberhard, former director of the Building Research Board, directed the overall effort; he was assisted by Gretchen Bank.

Peter Smeallie, *Editor*
Building Research Board

Contents

I. UNITED STATES

Association Profiles 3
Corporate Profiles 14
University Profiles 30
      Architecture and Design, 30
      Engineering, 40
Federal Laboratory Profiles 58
Other Profiles 71

II. INTERNATIONAL

Argentina, 77
Australia, 77
Austria, 81
Belgium, 84
Brazil, 90
Canada, 91
Chile, 97
China, 98
Columbia, 99
Czechoslovakia, 100
Denmark, 102
Ecuador, 107
Egypt, 108
Federal Republic of Germany, 108
Finland, 120
France, 123
German Democratic Republic, 127

Ghana, 128
Greece, 128
Guatemala, 129
Hungary, 129
Iceland, 131
India, 132
Indonesia, 134
Iraq, 135
Ireland, 136
Israel, 137
Italy, 138
Jamaica, 142
Japan, 143
Jordan, 149
Kenya, 150
Korea, 151
The Netherlands, 152
New Zealand, 159
Nigeria, 163
Norway, 163
Pakistan, 166
Philippines, 167
Poland, 168
Portugal, 171
Romania, 173
Singapore, 175
South Africa, 176
Spain, 178
Sweden, 180
Switzerland, 187
Tanzania, 191
Turkey, 191
USSR, 192
United Kingdom, 194
Venezuela, 210
Yugoslavia, 211

INDEX OF ORGANIZATIONS 213

INDEX OF SUBJECTS 221

An International Directory of Building Research Organizations

I

UNITED STATES

ASSOCIATION PROFILES

ACEC RESEARCH & MANAGEMENT FOUNDATION (ACEC/RMF)

1015 Fifteenth Street, N.W.
Washington, DC 20005
202/347-7474

CONTACT: Jack R. Warner, Vice President, Program Development

MISSION AND FOCUS OF RESEARCH: The foundation (ACEC/RMF) is an independent, nonprofit research and educational organization created by the American Consulting Engineers Council (ACEC) in 1974 to serve the public interest in the disciplines practiced by consulting engineers. The foundation's affiliation with ACEC ensures that its efforts are responsive to consulting engineers' needs and that ACEC/RMF research findings have accessible channels to design professionals and the public. The foundation helps the consulting engineering community by performing research and analyses to provide a factual basis for practice, educational tools, and policy-making. ACEC/RMF involves engineers in studies designed to transfer research findings into the practice environment. Most projects are developed with the assistance of an advisory committee of leading engineers and academics. Recent research projects have included building energy design, active solar energy design tools, R&D planning, seismic studies, energy efficiency, and passive solar energy design tools.

ACEC/RMF's annual budget of approximately $1.5 million is derived mostly from federal grants and contracts. The ACEC Scholarship Program, which awards thousands of dollars each year to engineering students, is supported by an endowment fund of ACEC/RMF. The foundation relies on advisory panels and consultants from the building industry to accomplish its work.

DISTINCTIVE ATTRIBUTES: Development of a data base of the standards referenced in codes and government design guides. Analysis of new flexible gas piping systems for buildings. Publication and research on retrofitting existing buildings for energy efficiency. Publication of studies on improving the efficiency of industrial processes. Energy analysis computer programs. Implementation of public agency design department management and operation reviews by A/E professionals. Revision of public agency design guide criteria. Comprehensive analysis risks to engineering firms in the hazardous waste field.

PUBLICATIONS: Reports, directories, and summaries. List available on request.

AMERICAN CONCRETE INSTITUTE

22400 West Seven Mile Road
Detroit, MI 48219
313/532-2600

CONTACT: George F. Leyh,
Executive Vice President

MISSION AND FOCUS OF RESEARCH: The institute works to further engineering and technical education, scientific investigation and research, and development of standards for the design and construction of concrete structures. It gathers, correlates, and disseminates information for the improvement of the design, construction, manufacture, use, and maintenance of concrete products and structures. The institute promotes improved technology technical competence, and good design and construction practices.

PUBLICATIONS: Concrete International: Design and Construction; Concrete Abstracts; ACI Materials Journal; ACI Structural Journal.

AMERICAN COUNCIL OF INDEPENDENT LABORATORIES

1725 K Street, N.W., Suite 412
Washington DC 20006
202/887-5872

CONTACT: Joseph F. O'Neil, Executive Director

MISSION AND FOCUS OF RESEARCH: The council is an association of independent testing, inspection, analytical, and research and development laboratories for clients in industry, commerce, and government.

AMERICAN INSTITUTE OF ARCHITECTS (AIA)/ACSA

1735 New York Avenue, N.W.
Washington, DC 20006
202/626-7300 AIA
202/785-2324 ACSA

CONTACT: Richard McCommons, Executive Director,
Association of Collegiate
Schools of Architecture

MISSION AND FOCUS OF RESEARCH: The AIA, which previously conducted and sponsored research through the AIA Foundation, formed a new research structure in 1986 with the Association of Collegiate Schools of Architecture (ACSA). The foundation continues as a public outreach organization. An advisory council of the AIA and ACSA sets priorities for a research agenda and promotes funding. Topics include energy; contracts and documents; and educational facilities.

AMERICAN INSTITUTE OF STEEL CONSTRUCTION (AISC)

The Wrigley Building
400 North Michigan Avenue
Chicago, IL 60611
312/670-2400

CONTACT: Nestor Iwankiw, AISC Director, Research and Codes
Larry Kloiber, Chairman, Committee on Research

MISSION AND FOCUS OF RESEARCH: AISC represents the U.S. fabricated steel industry. Research is aimed at supporting the use of fabricated structural steel through the sponsoring of projects selected by the AISC's Committee on Research. Recent work has included the study of large bracing connections, composite beam web openings, flush end-plate connections (with the Metal Building Manufacturers Association), composite simi-rigid connections, and electroslag welding. Other cooperative research priorities have been established within broad categories, for example, computer-aided design (CAD), steel fire resistance load and resistance factor design, seismic design and total building systems. AISC receives ongoing federal government support, from the National Science Foundation (NSF), National Cooperative Highway Research Program (NCHRP), and the Federal Highway Administration (FHWA) for steel research. The institute's own research budget is about $100,000.

PUBLICATIONS: *AISC Manual of Steel Construction, AISC Engineering Journal,* and other textbooks and publications.

AMERICAN IRON & STEEL INSTITUTE (AISI)

133 15th Street, N.W., Suite 300
Washington, DC 20005
202/452-7190

CONTACT: Albert C. Kuentz, Program Manager, Research

MISSION AND FOCUS OF RESEARCH: The institute sponsors research in steelmaking and steel applications including heavy construction, sheet steel, metal plate, steel framing, and other building-related topics. Funding is frequently collaborative or cooperative with other associations, industry and the federal government. The budget for building-related work is approximately $1 million.

Recent studies in steel construction include studies of structural reliability, autostress design of steel bridges, ultimate strength of composite plate girders, eccentrically-braced steel frames in earthquakes, structural design of cold-formed steel walls and columns, response of steel structure to fire.

PUBLICATIONS: Manuals, handbooks, and research reports.

AMERICAN PLYWOOD ASSOCIATION

Research Center
7011 South 19th Street
P.O. Box 11700
Tacoma, WA 98411
206/565-6600

CONTACT: William T. Robison, President,
American Plywood Association
Thomas R. Flint, Director,
Technical Services Division
Michael R. O'Halloran, Assistant Director
Technical Services Division

MISSION AND FOCUS OF RESEARCH: Market-oriented research aimed at solving building problems. Examples include the glued-floor system, permanent wood foundation, and diaphragm and shear wall design. APA's Technical Division also develops information leading to industry-adopted product and performance standards.

DISTINCTIVE ATTRIBUTES: A 37,000-square-foot research center. Materials and structural systems testing capability. Real-time computer system. A technical staff of more than 20 engineers and scientists.

PUBLICATIONS: More than 300 titles, ranging from do-it-yourself publications to the *Panel Specification Series*.

AMERICAN SOCIETY OF CIVIL ENGINEERS

345 East 47th Street
New York, NY 10017
212/705-7496

CONTACT: Edward Kippel, Manager, Technical Services

MISSION AND FOCUS OF RESEARCH: The society's purpose is to stimulate research in civil engineering. The Society's Technical Council on Research coordinates research proposals and interests of the council's 22 advisory committees. The society does not fund research, although it will provide administration of research funding. Current research interests include quality assurance in constructed projects and structural plastics.

AMERICAN SOCIETY OF HEATING, REFRIGERATING AND AIR-CONDITIONING ENGINEERS (ASHRAE)

1791 Tullie Circle, N.E.
Atlanta, GA 30329
404/636-8400

CONTACT: William W. Seaton, Manager of Research

MISSION AND FOCUS OF RESEARCH: ASHRAE's purpose is to advance the arts and sciences of heating, ventilation, refrigeration, air-conditioning, and related human factors. The Research and Technical Committee solicits proposals on research topics from universities and other agencies interested in cooperative research. Research topics of increasing interest include indoor air quality problems, system dynamics studies, building responses to climatic changes, and the effects of moisture and humidity on buildings. The 1987-88 research budget was $1.38 million, supported by individuals, chapters, and industry. The society has also conducted research for federal agencies, most recently the Department of Energy.

PUBLICATION: *ASHRAE Research Journal.*

AMERICAN SOCIETY OF PLUMBING ENGINEERS RESEARCH FOUNDATION

3617 Thousand Oaks Blvd.
Suite 210
West Lake Village, CA 913623
805/495-7120

CONTACT: John S. Shaw, Executive Director

MISSION AND FOCUS OF RESEARCH: The foundation's purpose is to coordinate, sponsor and develop funding for research in plumbing systems. It is currently developing a long-range agenda of research related to future water demand in commercial and residential buildings. Important issues being considered are the potential for safely reducing standard pipe sizes for supply and drainage, and supply and demand in large buildings. Using its long-range plan, the foundation will send out requests for proposals to universities and commercial laboratories. Its Advisory Council will support the research with contributions from ASPE's 4,500 members and from plumbing manufacturers. Funding from federal, state and local governments will also be considered.

THE ASPHALT INSTITUTE RESEARCH CENTER

Asphalt Institute Building
College Park, MD 20740
301/277-4258

CONTACT: Gerald S. Triplett, President
Vyt P. Puzinauskas, Director of Research
Edward T. Harrigan, Assistant Director of Research

MISSION AND FOCUS OF RESEARCH: The Research Center's focus is to improve current products and to develop new uses for bitumen and asphalt. The institute's laboratories on the campus of the University of Maryland do research and testing of materials used in road building, waterproofing, roofing, mastics, and other applications. A chemical laboratory is used to measure the physical properties of building materials; a physical laboratory is used for making mixtures used in paving. Both laboratories are equipped to do testing. The institute's 55 members are companies that manufacture asphalt products from crude petroleum. Its budget is about $4 million, of which about a third supports research. About 75 percent of the research effort is in the development of paving materials, and the remainder concerns structures. Three professionals and three technicians staff the institute's laboratories. The corporate members and the institute occasionally form cooperative research arrangements with universities.

PUBLICATIONS: Technical manuals, research reports, and audio-visual materials.

**ASTM
(formerly the American
Society for Testing
and Materials)**

1916 Race Street,
Philadelphia, PA 19103
215/299-5473

CONTACT: Spencer Everhardt, Information Resources

MISSION AND FOCUS OF RESEARCH: The ASTM establishes voluntary consensus standards for materials, products, systems, and services.

**THE BRICK INSTITUTE
OF AMERICA**

11490 Commerce Park Drive
Reston, VA 22090
703/620-0010

CONTACT: John Grogan, Executive Director
8601 Dunwoody Place, Suite 507
Atlanta, GA 30338
404/993-9714

MISSION AND FOCUS OF RESEARCH: The institute's programs address various aspects of brick masonry, including structural problems, water penetration resistance, and performance. The institute is also interested in the education of the design profession in the use of brick in the construction industry. Research programs are contracted out to universities.

**BUILDING OWNERS
& MANAGERS
ASSOCIATION
INTERNATIONAL**

1250 I "Eye" Street, N.W.
Suite 200
Washington, DC 20005
202/289-7000

CONTACT: Alton J. Penz, Staff Vice President, Research

MISSION AND FOCUS OF RESEARCH: The association, through a committee process with staff, studies needs and issues of importance to owners and managers of office buildings. It also conducts studies and keeps abreast of issues such as leasing, markets, codes and standards, income and operations expense, performance, energy operations, and rehabilitation/renovation. Other activities include the compilation of statistics, awards programs, and research programs.

DISTINCTIVE ATTRIBUTES: Oriented to empirical research on economic, operational, and institutional behavior issues. Publishes the most comprehensive annual survey available of income, expenses and operating characteristics for office buildings in North America.

PUBLICATIONS: *North American Office Market Review* (semi-annual); *Experience Exchange Report* (annual); *Standard Method of Floor Measurement*; special technical reports as appropriate (e.g., asbestos guidelines, fire incidence surveys, operating methods surveys).

**CONSTRUCTION
TECHNOLOGY
LABORATORIES, INC. (CTL)**

5420 Old Orchard Road
Skokie, IL 60077-4321
312/965-7500

CONTACT: Walter E. Kunze, President

MISSION AND FOCUS OF RESEARCH: The principal research focus is on the properties of cement and its performance in concrete, but laboratory work includes all other construction materials. CTL is one of the largest research facilities in its field in the world. Structures and materials are tested for fire, wind, earthquake resistance, strength, durability, and performance in a variety of uses. CTL provides performance evaluation, inspection, laboratory and field testing and product development and rehabilitation on a contract basis. The laboratory deals with all construction processes and materials, emphasizing cement and concrete technology.

DISTINCTIVE ATTRIBUTES: CTL has five buildings on a 22-acre site, with the world's largest multiaxial test facility, 50 million pounds capacity; facilities to test large-diameter pipe up to 700 pounds per square inch; a large, environmentally controlled test area; three firetesting furnaces; a calibrated hotbox; guarded hot plate; closed-circuit ball mill; computer-enhanced X-ray spectrometer; and pilot rotary kilns.

PUBLICATIONS: Journals, bulletins, and periodicals published by the Portland Cement Association and the CTL. An index is available.

EDISON ELECTRIC INSTITUTE

1111 19th Street, N.W.
Washington, DC 20036
202/778-6400

CONTACT: Norman Rubinstein, Director of Information and Publishing Services

MISSION AND FOCUS OF RESEARCH: The institute helps electric companies generate and distribute electrical energy at the lowest possible prices consistent with safe and reliable service. It promotes scientific research to meet people's needs through environmentally acceptable means, and it makes available information of importance to consumers and the industry.

ELECTRIC POWER RESEARCH INSTITUTE

3412 Hillview Avenue
Palo Alto, CA 94304
415/855-2000

CONTACT: Arvo Lannus, Program Manager, Residential and Commercial Energy Utilization Department
Mary Panke, Information Specialist, EPRI, 1800 Massachusetts Avenue, N.W., Washington, DC 20036
202/872-9222

MISSION AND FOCUS OF RESEARCH: In building-related areas, EPRI's research aims are to increase the efficiency and cost-effectiveness of electricity utilization, to improve end-use productivity, and to understand and control environmental, safety, and health effects of electricity use. The Residential and Commercial Program has a budget of $9 million, which is divided among research programs in heating and cooling, water heating, building envelope and indoor environmental control, thermal storage, lighting, refrigeration, appliances (including commercial cooking equipment), load control, automation, and metering. Related research is conducted in the demand-side planning program on customer response, marketing, and demand-side program planning methods and information development. All research is carried out by outside contractors.

PUBLICATIONS: *1987-1989 Research & Development Program Plan*, technical reports, *EPRI Journal*

GAS RESEARCH INSTITUTE

8600 West Bryn Mawr Avenue
Chicago, IL 60631
312/399-8100

CONTACT: James R. Brodrick,
Manager, Residential/Commercial
Technology Analysis

MISSION AND FOCUS OF RESEARCH: GRI plans, manages, and develops financing for a cooperative research and development program in the supply, transport, storage, and end use of gaseous fuels for the mutual benefit of the gas industry and its present and future customers. This institute is developing new and improved technologies that maximize the value of gas energy services while minimizing the cost of supplying and delivering gaseous fuels as the most effective way to serve the mutual interests of both the industry and its customers. To accomplish this program, GRI has a staff of 240 at its headquarters office and a budget for contract research of about $150 million per year.

The R&D program on end-use technologies develops improved gas-using equipment that is more effective, meets environmental standards, and offers competitive consumer costs at equivalent or higher quality of service compared with alternative energy service options. Research focuses on technologies such as heating systems, cooling systems, heat pumps, prime movers, appliances, fuel cells and energy cogeneration systems, and building systems for both residential and commercial sectors. To provide guidance for the R&D of end-use gas technologies, GRI sponsors building-related research projects. For example, detailed characterizations of the existing building stock are under way with Lawrence Berkeley Laboratory on multifamily residential housing and with Battelle Pacific Northwest Laboratories on office buildings. To field-test gas technologies, GRI is involved with several research test houses for a range of technical projects.

PUBLICATIONS: *Gas Research Institute 1986-1990 Research & Development Plan, Gas Research Institute Digest,* various technical reports, and staff papers.

INTELLIGENT BUILDINGS INSTITUTE

2101 L Street, N.W., Suite 300
Washington, DC 20037
202/457-8477

CONTACT: Richard H. Geissler, Executive Director
Jan Goebel, Associate Executive Director

MISSION AND FOCUS OF RESEARCH: The institute is a professional association serving all sectors of the intelligent buildings community. Activities include market information, training certification program's regulatory action/reaction, guidelines and standards, government advocacy, and joint research and development. Support for the institute comes from producers of hardware, software and systems, designers, consultants, architects, contractors, distributors, building owners, and managers.

DISTINCTIVE ATTRIBUTES: Multicompany membership focused on development of intelligent buildings.

LIGHTING RESEARCH INSTITUTE

345 East 47th Street, 9th Floor
New York, NY 10017
212/705-7511

CONTACT: Richard L. Vincent, Director of Development and Programs

MISSION AND FOCUS OF RESEARCH: The institute funds research on lighting, including photobiology; the health implications of artificial lighting; the relationship between lighting and vision; the integration of lighting, artificial and natural, in systems applications; and psychological aspects of human response to light, such as a recent study of how lighting influences the movement of people to exits in an emergency. The institute undertakes cooperative projects with other organizations, notably the Electric Power Research Institute, the U.S. Department of Energy, the National Electric Manufacturers Association, and the New York State Energy Research & Development Authority. The institute was founded in 1982 by the Illuminating Engineering Society of North America. It sponsors about $100,000 worth of research annually.

METAL BUILDING MANUFACTURERS ASSOCIATION

1230 Keith Building
Cleveland, OH 44115
216/241-7333

CONTACT: Gilliam Harris, Director of Research

MISSION AND FOCUS OF RESEARCH: The association's sponsored research focuses on the reaction of metal plates and steel buildings to different load conditions such as snow, wind, and seismic loading. Research is done on steel building design and analysis. Work is contracted out to universities. Several cooperatively funded projects are carried out with other steel-related organizations. The research budget is about $100,000.

NATIONAL ASSOCIATION OF HOME BUILDERS RESEARCH FOUNDATION

627 South Lawn
P.O. Box 1627C3
Rockville, MD 20850
301/762-4200

CONTACT: David J. MacFadyen,
Executive Vice President

MISSION AND FOCUS OF RESEARCH: The NAHB Research Foundation focuses on the introduction of significant new developments in building products and manufacturing practices. The foundation is a subsidiary corporation of the National Association of Home Builders. Research is conducted in five divisions: laboratory services, building systems, industrial engineering, economic studies, and special services. The foundation's laboratory is equipped with a 200,000-pound universal testing machine, air filtration measurement equipment, air quality instrumentation, insulation testing apparatus, weather simulators, and controlled temperature-humidity chambers. The foundation conducts design and testing of light-framed structures, construction of research houses, cost-effectiveness studies, and economic analyses. The foundation operates on a budget of $3 million, with a staff of approximately 45, including engineers, economists, architects, urban planners, and test technicians.

DISTINCTIVE ATTRIBUTES: The Smart House project, a consortium of 30 businesses formed to develop a new system for residential wiring.

PUBLICATIONS: A technology transfer program coordinates publications. Information, publications (through the NAHB bookstore), and surveys of various aspects of the building industry are available.

NATIONAL CONCRETE MASONRY ASSOCIATION

2302 Horse Pen Road
P.O. Box 781C3
Herndon, VA 22070
703/435-4900

CONTACT: Edwin G. Hedstrom, Director,
Research and Development

MISSION AND FOCUS OF RESEARCH: The association's laboratory performs structural research and testing on a variety of aspects of concrete masonry for members and on a contract basis.

DISTINCTIVE ATTRIBUTES: Facilities include a 2-million-pound test frame capable of testing walls 20-foot x 6-foot; a lateral test frame with air bag; water permanence test chambers; and environmental chambers for freeze testing, as well as a variety of other testing equipment.

CORPORATE PROFILES

ARMSTRONG WORLD INDUSTRIES, INC.

P.O. Box 3511
Lancaster, PA 17604
717/397-0611

CONTACT: Joseph E. Hennessey,
Assistant Director of Research

MISSION AND FOCUS OF RESEARCH: Armstrong concentrates on development of new fibrous compositions for ceilings, integrated lighting and ventilating ceiling systems, and processes for fiberboard decoration. In addition, the company conducts research into improving existing products such as fire-resistant and fire-retardant material for flooring, carpeting, ceilings and walls, thermal insulation, acoustical wall products, and furniture. The Armstrong Technical Center employs more than 500 technical and support personnel with a 1985 budget of $46.8 million.

DISTINCTIVE ATTRIBUTES: Extensive pilot plant facilities.

AUTODESK, INC.

2320 Marinship Way
Sausalito, CA 94965
415/332-2344

CONTACT: J. Clifford Gauntlett,
AEC Technology Research

MISSION AND FOCUS OF RESEARCH: Autodesk develops low-cost CAD/CAM software, portable across a variety of PC and engineering workstations, providing design, 3-D visualization and data base standards for architectural uses, building construction, and facilities management function.

AUTO-TROL CORPORATION

12500 N. Washington Street
P.O. Box 33815
Denver, CO 80233
303/452-4919

CONTACT: David Weisberg

MISSION AND FOCUS OF RESEARCH: The company's major focus is on the development of turnkey computer-assisted design system software for structural design and analysis, as well as traditional engineering drawing and preparation. The corporate research and development group consists of 125 individuals. Contract research in CAD/CAM software is possible.

**BASF
CORPORATE RESEARCH AND
DEVELOPMENT CENTER**

1419 Biddle Avenue
Wyandotte, MI 48192
313/246-6100

CONTACT: Gerhard Paul, Vice President, Corporate R&D

Wolfgang Martin, Vice President, R&D,
 ASF Corporation Fibers Division,
 P.O. Drawer D,
 Williamsburg, VA 23187
 804/887-6000

MISSION AND FOCUS OF RESEARCH: BASF Corporation is a subsidiary of an international chemical and plastics company, BASF AG, which has its major technical center in West Germany. The corporate R&D center studies chemicals, plastics, foam materials, and applications of these, including their use in buildings and as composites. The corporation's Fibers Division in Williamsburg, Virginia, studies the properties of fibers, processes for production, and development of products. The company is interested in applications of fiber in construction materials instead of asbestos, such as its use in buildings as carpeting.

**BOLT, BERANEK, AND
NEWMAN LABORATORIES, INC.**

10 Moulton Street
Cambridge, MA 02138
617/491-1850

CONTACT: Frank E. Heart, Senior Vice President,
 Director of Computer and
 Information Sciences
Francis J. Jackson,
 Senior Vice President,
 Director, Physical Sciences

MISSION AND FOCUS OF RESEARCH: BBN Laboratories, Inc., is the research and development arm of Bolt, Beranek, and Newman, Inc. The subsidiary offers contract research services with an emphasis in computer, information, and physical sciences. Recent projects include work in multiprocessors; intelligent systems including a pharmaceutical manufacturing process design tool, a molecular modeling system, an avionics training aid, a computer assembly configuration management program, and a network topology and route assessment system; natural language and knowledge representation; and environmental and architectural acoustics.

DISTINCTIVE ATTRIBUTES: Uses small-scale test models and simulation.

PUBLICATIONS: Internal papers and trade publications.

CALMA COMPANY

501 Sycamore Drive
Milpitas, CA 95035
408/434-4000

CONTACT: Malcolm Davies

MISSION AND FOCUS OF RESEARCH: The primary business of CALMA is application, enhancement, and improvement of AEC Computer Assisted Design Systems. The company employs 400 people involved in research and development and had a research budget of $45 million in 1987.

CATERPILLAR, INC.

Technical Center-Bldg. A
100 N.E. Adams St.
P.O. Box 1875
Peoria, IL 61656-1875
309/578-6777

CONTACT: C. E. Grawey, Director of Research

MISSION AND FOCUS OF RESEARCH: Caterpillar, Inc., addresses the improvement of products primarily with respect to diesel engines, earth-moving equipment, stand-by power units and diesel electric sets. Of 3,000 employees at the technical center, 300 are research professionals, with approximately 75 mechanical engineers, 65 working on metallurgics and ceramics, 100 working on nonmetallics such as paints and plastics, and 75 working on new vehicle concepts, CAD/CAM projects, model building, and finite elements.

CONTROLLED DEMOLITION INTERNATIONAL

2737 Merryman Mill Road
Phoenix, MD 21131
301/667-6610

CONTACT: Mark Loizeaux, President

MISSION AND FOCUS OF RESEARCH: As an international contractor, the company has experience in all aspects of large building demolition using conventional and explosive demolition methods, as well as standard and specialty debris-handling techniques. Data gathering is performed on all previous work with respect to forces needed to handle a variety of construction materials resistant to structural or material failure and the ability of materials to withstand pressures and failure of different kinds. Extensive blast data are also compiled.

DISTINCTIVE ATTRIBUTES: The firm has joined with a group of experts in antiterrorist design using computer-aided design (CAD) to develop a secure buildings program. Has international experience in rescue efforts, debris removal, and demolition following earthquake damage/natural disasters. Has experience with dismantlement of nuclear facilities.

CORNING GLASS WORKS

Sullivan Park
Building Code FR-2-10
Corning, NY 14831
607/974-3331

CONTACT: Dr. William R. Prindle,
Associate Director, Research,
Development, and Engineering

MISSION AND FOCUS OF RESEARCH: Current activities include development of synthetic glass material for use in architectural fascia, including glass and ceramic shingles. Other efforts are directed at the use of glass-ceramics, which are impact-resistant, durable, and relatively inexpensive and which can be used for walls and flooring. Corning has taken on a range of projects that include development of missile radomes, infrared transmitting glasses, telescope mirrors, optical waveguide fibers, reinforced glass-ceramic composites, and extruded cellular ceramics for heat transfer combustor applications.

Formed in 1908, the corporate research and development center is a major inorganic materials research center engaged in programs focused on glass, glass-ceramics, technical ceramics, and other associated materials. The laboratories, in addition to research and development activities directed toward the internal needs of Corning Glass Works, have expanded their scope to include contract services for government and industrial sponsors.

With a total staff of 750 and an annual budget of $50 million, the research and development center provides its services either as a prime contractor or as a subcontractor to government agencies and industrial firms.

DISTINCTIVE ATTRIBUTES: Melting and forming of large glass articles and ceramics. Capability to form advanced ceramic material reinforced with silicon carbide or carbon fibers.

PUBLICATIONS: Corning RESEARCH: Articles by Corning Glass Works Scientists (annual), internal papers, and professional journals.

DEERE AND COMPANY Technical Center
John Deere Road
Moline, Il 61265
309/752-8000

CONTACT: Brian Alm, Manager, Media Relations
Mack Klimpler, Manager, Engineering (Industrial Products),
Dubuque, Iowa
319/589-5287
Robert Wismer, Director, Technical Center

MISSION AND FOCUS OF RESEARCH: The budget for research and development has averaged just over $200 million for the past 5 years, emphasizing the coordination of product development and manufacturing of heavy equipment in order to improve product quality, manufacturing production, and cost efficiencies. Research is diversified throughout the company, with each manufacturing plant containing an engineering department, while all draw on the corporate R&D center for coordination and scientific application assistance. Research is also undertaken in chemical and physical sciences.

DISTINCTIVE ATTRIBUTES: Product development and design in heavy equipment industry.

EVERETT I. BROWN COMPANY, ARCHITECTS AND ENGINEERS 941 North Meridian Street
Indianapolis, IN 46204
317/266-0033

CONTACT: Joseph Brown,
Managing Partner and Director
John W. Howard, Marketing Representative

MISSION AND FOCUS OF RESEARCH: The company is a full service A/E firm with in-house architectural capability and mechanical, civil, structural, and electrical engineering disciplines. Among the first in the United States to employ computer-aided design as an operational tool, the firm now has 54 work stations, 11 CPUs and more than 7,000 megabytes of storage capacity supplemented with PCs, printers, and alphanumeric input devices. The firm has its own in-house software developers.

The company promotes the design of secure buildings through the formation of interdisciplinary teams of architects/engineers and security specialists whose assessment and performance abilities are enhanced through the sophisticated use of analytic capabilities available in computer-aided design technology.

DISTINCTIVE ATTRIBUTES: Experience in CAD systems.

FACTORY MUTUAL ENGINEERING AND RESEARCH

1151 Boston-Providence Turnpike
P.O. Box 9102
Norwood, MA 02062
617/762-4300, Ext. 1405

CONTACT: Madeleine Andersson, Director, Public Relations

MISSION AND FOCUS OF RESEARCH: The organization's research supports its loss control services and is mainly concerned with fire, the nature of combustion, fire hazard characteristics of materials, fire prevention and control, and testing and loss analysis. Explosions, sprinkler systems, convection, and mine fires are also studied.

DISTINCTIVE ATTRIBUTES: The organization's large-scale fire test facilities in West Gloucester, Rhode Island, can be used to set up and study full-size fires under controlled conditions.

FORD MOTOR COMPANY, GLASS DIVISION

Product Research & Development
15000 N. Commerce Drive
Dearborn, MI 48120
313/845-4790

CONTACT: Glenn Stinson, Support Laboratories Manager

MISSION AND FOCUS OF RESEARCH: The division, with 45 engineers and a budget of several million dollars, draws on the support of Ford Motor Science Labs, which has a billion-dollar-plus budget and extensive facilities. Research is conducted in commercial, architectural, and automotive glass, primarily in three specific areas: coated products, both pyrolytic and vacuum "sputter" coatings; curtainwall construction; and low-emissivity or heat-reflecting glass. The primary focus of these products is energy conservation, architectural economics, and aesthetics. Advanced research is being conducted in areas such as electroluminescence, photovoltaics, liquid crystal switching, electrochromics, and decorated products.

DISTINCTIVE ATTRIBUTES: Electron microscope for analyzing composition of particles and coatings, heat-up facilities, vacuum sputter coating machines, and computer X-ray.

PUBLICATIONS: Trade and professional journals.

GUARDIAN INDUSTRIES

Architectural Glass Division
14600 Romaine Road
Carleton, MI 48117
313/962-2252

CONTACT: Ray Nalepka, Director, Research and Development

MISSION AND FOCUS OF RESEARCH: Guardian Industries' research focuses on the development of a wide variety of architectural glasses, including float glass for building exteriors; high-performance reflective glasses with a variety of coatings, including low emissivity; reflective spandrel glass used for building exteriors; insulating glass; acoustical window systems; and laminated glass. Laboratories include equipment for stress testing, emissivity properties, and glass composition.

DISTINCTIVE ATTRIBUTES: Refective glasses.

HONEYWELL INC.

Honeywell Plaza
Minneapolis, MN 55408
612/870-5200

CONTACT:
Susan Eich, Corporate Public Relations, Ext. 6730
Jim Olson, Vice President of Engineering Building Controls Division, Ext. 6684
Gene Chodash, Director of Engineering, Commercial Building Groups, Ext. 5350
Roger Feulner, Vice President and General Manager, Corporate Systems Development Division, Ext. 6887
Paul Peterson, Director, Physical Sciences Center, Ext. 4307
Mary Nelson, Protective Services Division, Ext. 2862

MISSION AND FOCUS OF RESEARCH: Honeywell Inc. divides its research into five divisions: Building Controls Division, Commercial Buildings Group, Corporate Systems Development Division, Physical Sciences Center, and Protection Services Division.

The Building Controls Division produces heating, ventilation, and air conditioning controls; lighting controls and building management systems for commercial buildings; burner and boiler combustion controls; and water controls. In addition, the division provides its customers with indoor air quality diagnostic services.

The Commercial Buildings Group comprises two divisions: Building Services and Building Systems. Building Services Division provides operational support, maintenance, communication, and energy retrofit systems and services under contract to owners and operators of commercial, industrial, and institutional buildings. Building Systems Division produces comfort controls and building management, fire-alarm, and security systems for commercial, industrial, and institutional buildings. The Integrated Building Center, within the division, provides integrated building control and communication systems for factories and plants, office buildings, correctional institutions, health-care institutions, hotels, retail stores, and schools and universities.

The Corporate Systems Development Division provides many products and services for building markets, such as systems engineering for integration of control, communications, and computer systems for commercial buildings and industry. The division also designs and develops multipurpose application processors. In addition, it integrates building wiring, designs special workstations, and creates communication protocols.

The Physical Sciences Center was formed to aid in the research and development of sensing and control technologies for automation.

The Protection Services Division provides security services for homes and commercial buildings. This includes burglary and fire protection, access control, and 24-hour monitoring by the division's customer service centers.

Research and development spending in 1986 amounted to $705 million, with $247 million, or 4.6% percent of sales, being Honeywell-funded and the rest customer- funded. Honeywell is a contributor to the Smart House Consortium of the National Association of Home Builders and the Electronic Industries Association.

JOHNSON CONTROLS, INC.

5757 North Green Bay Avenue
P.O. Box 591
Milwaukee, WI 53201
414/228-1200

CONTACT: George Jacobi, Vice President for Technology
Dennis Miller, Manager, Controls Research

MISSION AND FOCUS OF RESEARCH: Johnson Controls focuses on the integration of building controls, lighting, heating, ventilation, air conditioning, fire security, and communications and wiring.

KAWNEER COMPANY, INC.   555 Guthridge Court
Norcross, GA 30092
404/449-5555

CONTACT: Robert D. Foster, Director, Research and Development

MISSION AND FOCUS OF RESEARCH: Kawneer conducts applied research for the development of new products. The firm is a producer of windows, curtain walls, doors, entrance framing, hardware, exterior storefronts, and panel products with fire-resistant properties.

DISTINCTIVE ATTRIBUTES: Chemical anodizing labs, and static and thermal chambers.

KIDDE AUTOMATED SYSTEMS   35 Sharon Drive
West Lake, OH 44105
216/871-9900

CONTACT: Richard Evans, Director, Research and Development
Nick Martello, Product Specialist

MISSION AND FOCUS OF RESEARCH: Product development and the improvement of access, fire, and security and building automation control systems are major research areas at Kidde. The company also specializes in communications between control panels and system computers, multiple area networks, and transmit techniques as interfaces and system integrators. The company is currently testing use of fiber optics in control systems.

KOPPERS COMPANY, INC.   436 Seventh Avenue
Pittsburgh, PA 15219
412/227-2307

CONTACT: Charles P. Dorsey, Vice President, Science and Technology

MISSION AND FOCUS OF RESEARCH: Koppers focuses on the exploration of advanced technologies in construction materials and chemical products, as well as the development of new products and the improvement of manufacturing processes in these areas. The company sponsors external research at three universities and at high-technology R&D companies in which it has an interest. The company's main building product areas are roofing, roof insulation, phenolic foam insulation, waterproofing systems, roofing membranes, and asphalt and reinforced coatings, aggregates, asphalt concrete, etc.

LENNOX INDUSTRIES INC. Research Center
1600 Metrocrest Drive
Carrollton, TX 75006
214/245-2525

CONTACT: Dave Treadwell, Director of Engineering

MISSION AND FOCUS OF RESEARCH: Lennox emphasizes the testing and improvement of heating, ventilating, and air conditioning products in its research. The company is one of the largest furnace manufacturers in the United States and conducts research in conjunction with the American Gas Association, Gas Research Institute, and Battelle Memorial Institute.

The research and development group consists of 200 people, 64 of whom are engineers involved in product development, evaluation, design services (both manual and CAD/CAM), engineering, and rehabilitating components. Facilities include: complete model shop, fabricating of sheet metal, welding and machine shop, packaging lab, electric heat testing lab, a sound testing lab, air handling lab, metallurgy and chemical labs, and test facilities for cooling performance. Lennox is currently involved in a power control project with the Salt River Project, an electric utility company in Arizona. The applied research group also participates in outside contract work.

NATIONAL GYPSUM   Gold Bond Research Center
1650 Military Road
Buffalo, NY 14217
716/873-9750

CONTACT: Terry Williamson, Vice President

MISSION AND FOCUS OF RESEARCH: Primary activities are centered on wallboard products such as joint compounds, paper, and wallboard; construction systems for the application of wallboard; steel products for attaching wallboard; and exterior vinyl sidings. The company also evaluates new processes for making gypsum board. Additionally, the company tests fire-resistant products for outside concerns. Research activity is supported by a staff of 60 technical people.

DISTINCTIVE ATTRIBUTES: Fire testing utilizing a full-size horizontal and vertical wall furnace and a tunnel furnace to test fire spread. Also has an acoustical testing facility.

PUBLICATIONS: Internal reports and trade publications.

OWENS-CORNING FIBERGLAS Technical Center
Granville, OH 43023
614/587-0610

>   CONTACT: Frank Beumel, Laboratory Director, Construction Products Group

>   **MISSION AND FOCUS OF RESEARCH:** The company produces materials designed to achieve thermal or acoustical insulation; electrical resistance, ballistic properties; dimensional stability; high strength with reduced weight and greater flexibility or stiffness.

>   Fundamental structures of materials are studied, as well as mechanical, thermal, and chemical properties. New techniques are being designed to develop materials based on the scientific disciplines of glass and polymer synthesis, the physics of fiber forming and the engineered processing of glass fibers and resins. The 1987 science and technology department budget of $45 million supports a staff of about 500 technical and support personnel at the Granville, Ohio, Technical Center, and several smaller laboratories at plant sites.

>   **DISTINCTIVE ATTRIBUTES:** Fatigue lab for testing long-term durability; and fire testing labs for tests of safety of construction materials and interior fabrics, thermal properties.

NORTH AMERICAN PHILIPS LIGHTING CORPORATION 200 Franklin Square Drive
CN 6800
Somerset, NJ 08873

>   CONTACT: Rob Peters, Vice President, Corporate Research and Development
>   Robert McCulley, Director Product Management
>   Paul Rorer, Manager, Application Engineering

>   **MISSION AND FOCUS OF RESEARCH:** Philips pursues new and improved product development research with respect to the lighting of most internal and external building spaces with a variety of light fixtures and light sources. At the corporate research facility in Holland, state-of-the-art research and application engineering are being conducted on aspects of lighting not generally accepted in U.S. markets, such as reflector panels, low-energy light sources, new types of light fixtures, and increasing lamp life.

>   The U.S. research is concentrated on increased energy efficiency and longer life of light sources; identification of new light sources, including new uses and applications for low-pressure

sodium; high-intensity discharge lamps; compact fluorescents; and tungsten halogen lamps. Development of fixtures is also an important aspect of light generation and efficiency. The company also conducts research into photometrics and the use of reflectors and refractors.

Among a variety of research topics currently under consideration are workplace issues, including productivity and its relationship to light when standards are set for various task requirements, various ages of individuals, and the efficiency of the system. Included in this category is a move back to task lighting from wide-area lighting, which was pursued in response to energy efficiency.

DISTINCTIVE ATTRIBUTES: Efficiency of low-sodium lighting.

PUBLICATIONS: Extensive use of external and internal books, journals, guides, and newsletters.

PPG INDUSTRIES

One PPG Place
Pittsburgh, PA 15272
412/434-3131

PPG Glass R&D
Harmarville, PA 15238
412/665-8500

CONTACT: R. L. Scriven, Director, Glass Research and Development, Harmarville

MISSION AND FOCUS OF RESEARCH: PPG research emphasizes development and improvement of glass and coating products. Glass research is involved with energy efficient coatings for residential and commercial construction, impact resistance, and decorative architectural glass for interior and exterior use. Coatings research is focused on electrodeposition coatings, high-speed radiation-curable coatings, and heavy-duty industrial maintenance coatings.

SCHLAGE LOCK CO.

2401 Bayshore Blvd.
San Francisco, CA 94134
415/467-1100

CONTACT: Lee Tatara, Acting Director of Research

MISSION AND FOCUS OF RESEARCH: Schlage develops and improves products used in locking systems and builder's hardware. The company is involved in the Smart House Consortium of the National Association of Home Builders and the Electronics Industries Association.

It intends to concentrate research on electronic security systems, doubling its present level of effort from 14 to 28 people.

DISTINCTIVE ATTRIBUTES: Participates in joint ventures to develop security systems.

STEVEN WINTER ASSOCIATES, INC.

6100 Empire State Bldg.
New York, NY 10001
212/564-5800

170 Newport Center Drive
Newport Beach, CA 92660
714/760-1568

CONTACT: Steven Winter, President
Deane Evans, Principal

MISSION AND FOCUS OF RESEARCH: An architectural/engineering firm that provides extensive technical consulting services to the building industry, including design, structural engineering, mechanical engineering, value engineering, code compliance, energy modeling, construction supervision, asbestos assessments, energy audits, technical publications, CAD/CAM, cost analysis, technical research, hazards research, new industry development, and technology transfer. The firm provides these services to a wide range of clients, including developers; product manufacturers and suppliers; federal and state government agencies; industry associations and utilities; and design professionals and consultants.

DISTINCTIVE ATTRIBUTES: Expertise in prefabricated and industrialized housing, market analysis, and computer modeling.

PUBLICATIONS: *Passive Solar Construction Handbook; Greenhouses for Living; Energy-Efficient Lighting Design; Energy Conservation Standard for Federally Procured Housing; Affordable Housing Through Energy Conservation; Designing Affordable Housing; Home Building Cost Cuts; Suntempering in the Northeast; Building Value into Housing, 1980 Awards; Residential Steel Frame Construction.*

GTE SYLVANIA PRODUCTS CORPORATION

100 Endicott St.
Danvers, MA 01923
617/777-1900

CONTACT: John Waymouth, Director, Research and Development

MISSION AND FOCUS OF RESEARCH: GTE Sylvania focuses on the development of energy-efficient lights for indoor and outdoor application. Primary emphasis in recent years has been research in support of new products.

DISTINCTIVE ATTRIBUTES: Sylvania develops and markets tungsten-halogen lamps, fluorescent lamps, and high-intensity discharge lamps for all lighting specifications.

PUBLICATIONS: Publishes extensively in both trade and professional publications.

3M COMPANIES 3M Center
St. Paul, MN 55101
612/733-7177

    CONTACT: Tom Severeide, Executive Director,
                       Corporate Research Laboratories
                Gerry Kottong, Government Research Department

MISSION AND FOCUS OF RESEARCH: 3M is primarily concerned with improving and developing new products in existing lines including abrasives, adhesives, chemicals, tapes, and electrical products. Activities include work in damping tall buildings; calks and seals; fire retardation and flammables; reflective and efficient light fixtures; and a fire containment system consisting of "fire barrier" compound inside-wall material.

DISTINCTIVE ATTRIBUTES: An annual R&D budget of about $468 million comprises about 5.7-6.0 percent of sales. About 6,000 technical professionals are employed in R&D.

PUBLICATION: Trade literature.

USG CORPORATION Research and Development Center
700 North Highway 45
Libertyville, IL 60048
312/362-9797

    CONTACT: Mitch Ptasienski, Manager, Information Services

MISSION AND FOCUS OF RESEARCH: With more than 3,000 products, USG Corporation emphasizes research in improved and new products. General categories of products include gypsum board; joint treatment; building plasters; acoustical and insulating tiles; steel components such as studs, joists, corner beads, grids, and clips of various kinds; industrial plasters; fillers; lime; agricultural gypsum; and wood fiber items. Research is supported by 155 technical staff. In addition, Masonite Corporation and Donn Products maintain seperate research facilities.

UNITED TECHNOLOGIES United Technologies Building
Hartford, CT 06101
203/728-7000

    CONTACT: James W. Clark, Assistant Director of
                       Research for Engineering Operations
                United Technologies Research Center
                Silver Lane
                East Hartford, CT 06108
                203/727-7080

MISSION AND FOCUS OF RESEARCH: As a broad-based designer and manufacturer of high-technology products, United Technologies has a variety of products and research capabilities. Subsidiaries of the corporation include Otis Elevator, the world's largest elevator manufacturer, and Carrier Corporation, worldwide leading air-conditioner manufacturer. The United Technologies Research Center conducts basic and applied research across a broad spectrum of advanced technologies employing about 1,100 technical and support personnel. The Research Center performs near-, medium-, and long-term R&D programs in photonics (integrated optics, fiber-optics, lasers, electro-optics, nonlinear optics, acousto-optics); electronics (compound semiconductor materials and devices, microelectronic sensors, hybrid circuits, signal processing, microwave devices); and computer science (artificial intelligence, computer architectures, software), which can have application to integrated security systems, special alarm and detection systems, communications systems, and the detection and analysis of electromagnetic emissions.

DISTINCTIVE ATTRIBUTES: Otis recently completed a new 28-story test tower on a 15-acre site in Bristol, Connecticut. This new research facility represents an investment of about $15 million.

WESTINGHOUSE ELECTRIC CORPORATION

Research and Development Center
1310 Beulah Rd.
Pittsburgh, PA 15235
412/256-2800

CONTACT: Issac R. Barpel, Vice President, Research and Development
Doris A. Pollitt, Manager, Communications, 412/256-2721

MISSION AND FOCUS OF RESEARCH: The center responds to the needs of the corporation's operating divisions, provides access to expertise needed by the operating divisions, and transfers technology from the center to the operating divisions. The center consists of seven technical divisions: applied sciences, electronics technology, chemical sciences, solid state sciences, materials sciences, solid oxide technology, and engineering. Most recently, Westinghouse has pursued work on high-temperature solid oxide fuel cells for production of electricity as a total energy system for residences and industrial cogeneration systems. The center has a staff of 1,500, including 600 scientists and engineers.

W.R. GRACE & CO. Construction Products Division
62 Whitmore Ave.
Cambridge, MA 02140
617/876-1400

CONTACT: Barry Siadat, Director of New Business,
Development and Research
Jan Skalny, Director,
Construction Specialties Group
Washington Research Center,
W.R. Grace & Co., Columbia, Md.
301/531-4597

MISSION AND FOCUS OF RESEARCH: Research activities are focused on cement and concrete additives, thermal barriers to fire, moisture and noise transmission, durability of structural envelopes, roofing systems, and materials including concrete, gypsum, rubbers and plastics, and metals. Generally, the research program is concerned with system compatibility, performance of buildings, and fireproofing systems. The budget for research in construction products is several million dollars per year with a staff of over 100 technical and support persons. A corporate R&D center, the Washington Research Center, at Columbia, Maryland, includes a construction specialties group.

DISTINCTIVE ATTRIBUTES: Thermal measurement laboratory. Laboratory for cement and concrete testing.

PUBLICATIONS: Internal publications, journals, patents, and professional conferences.

UNIVERSITY PROFILES

Architecture and Design

CARNEGIE-MELLON UNIVERSITY

Center for Building Diagnostics
Institute of Building Sciences
Department of Architecture
Doherty Hall-Schenley Park
Pittsburgh, PA 15213
412/268-2350

CONTACT: Volker Hartkopf

MISSION AND FOCUS OF RESEARCH: The center emphasizes research in four major areas: (1) evaluation of building performance, including researching the area of total building performance, modeling individual areas of building performance such as light and energy, and indoor air quality; (2) computer-aided design, including three-dimensional models, expert systems, and building performance simulations; (3) the science of design; and (4) building development of Third World countries, including postdisaster shelter, predisaster planning, housing development in earthquake-prone areas, and improved technology introductions. The annual budget is $600,000 with a full-time staff of 12.

DISTINCTIVE ATTRIBUTES: Facilities include a CAD laboratory and monitoring equipment used in building performance analysis.

PUBLICATIONS: Building research series and professional journals.

UNIVERSITY OF CALIFORNIA

College of Environmental Design
Berkeley, CA 94720
415/642-0830

CONTACT: Richard Bender, Dean

MISSION AND FOCUS OF RESEARCH: The two related units are the Center for Environmental Design Research and the Institute of Urban and Regional Planning. Between the two they cover the areas of seismic hazards and building safety, Third World urban development,

Third World urban design and planning, fire, life safety, energy, information technology, accessibility for the elderly and handicapped, and building science and site diagnostics. The building science laboratory has a wind tunnel, an artificial sky, environmental chambers, computer graphics simulation, and environmental design simulation. The staff includes 20 senior faculty members, 50 graduate students in architecture, 10 in landscape architecture, and 10 in city planning. They have a budget that ranges from $1 million to $2 million depending on grants. Sources are federal, state and city governments; the university; and industry.

DISTINCTIVE ATTRIBUTES: A unique combination of environmental simulator, artificial sky, and wind tunnel.

PUBLICATIONS: Reports and journals.

UNIVERSITY OF COLORADO

College of Design and Planning
Campus Box 314
Boulder, CO 80309
303/492-7711

CONTACT: Louis Sauer

MISSION AND FOCUS OF RESEARCH: Areas of study within the College of Design and Planning include the nature of innovation in the homebuilding industry; the technology of the open-frame wood system; the nature and structure of mobile manufactured homes; programming computer design programs with hierarchically organized questions; the nature of housing preferences; the criteria for evaluation, programming and design of residential settings; the impact of housing choice and fertility; natural hazard mitigation; and behavioral aspects of design.

PUBLICATIONS: Independent reports and papers.

COLUMBIA UNIVERSITY

Center for Preservation Research
Graduate School of Architecture, Planning, and Preservation
400 Avery Hall
New York, NY 10027
212/280-3414

CONTACT: Frank G. Matero, Director

MISSION AND FOCUS OF RESEARCH: The center offers applied consulting and research services to address issues of architectural conservation as a combination of fine arts and architecture. Coursework involves analysis of historic building materials including concrete, stone, brick, terra cotta, decorative and painted finishes, and architectural metals. Conservation treatments are also performed. The staff consists of four principal researchers and research technicians. Facilities include a microscopy laboratory, two research laboratories, and a workshop, as well as access to Columbia's testing and library facilities.

DISTINCTIVE ATTRIBUTES: The only program and facility of its type and size dealing with architectural conservation.

FLORIDA A&M UNIVERSITY

Institute for Building Sciences
School of Architecture
Florida A&M University
Tallahassee, FL 32307
904/599-3244

CONTACT: Thomas Martineau, AIA, Director

MISSION AND FOCUS OF RESEARCH: The institute focuses on continuing education--seminars, workshops, conferences--on topics of interest to the building community. Both basic and applied research are conducted as well as single-client and multiclient sponsored research projects for private industry or public sector agencies. In the area of technical assistance, the institute conducts training programs for citizens of developing nations in construction methods, technology, and management. As a service to the domestic industry, the institute also collects, stores, and shares information on problems in construction, operation, maintenance, and rehabilitation of buildings of all types.

DISTINCTIVE ATTRIBUTES: Designated center of excellence in the Florida university system.

PUBLICATIONS: *The Intelligent Building Service of the Institute of Building Sciences; Building Construction Regulations in Florida*; various reports and papers on low-cost construction, architects in corporations, fixture standards for primary and secondary educational facilities, and turnkey programs for industrial and institutional laboratory facilities.

GEORGIA INSTITUTE OF TECHNOLOGY

Center for Architectural Conservation
College of Architecture
Atlanta, GA 30332
404/894-3390

CONTACT: John H. Myers, Director

MISSION AND FOCUS OF RESEARCH: The center has the capability to provide expert systems in conservation, thorough assessments of existing building conditions to determine deficiencies, cost estimates, and recommendations for conservation, rehabilitation, and maintenance. As part of this process, buildings are documented completely by means of videotaping or photography. Segments can be assembled in any fashion, and retrieval of this information is immediate. A typical building assessment includes analysis of roofs, foundations, materials, HVAC, fire safety, and the immediate site. All research is computer automated. Facilities include computers, video equipment, a mobile motor home for field investigation of buildings, and the use of the building materials laboratory. There are 8 people in the center, but staff size varies from 4 to 25, including faculty members and graduate research assistants. The budget varies from $400,000 to $1 million, depending on grants, with the Department of the Interior as the primary source.

HARVARD UNIVERSITY

Department of Architecture
Graduate School of Design
Cambridge, MA 02138
617/495-2294

CONTACT: Daniel Schodek

MISSION AND FOCUS OF RESEARCH: Research projects include CAD during the early phases of design, structural modeling, lighting, acoustics, three-dimensional modeling, and artificial intelligence (AI) systems. Other activities are site engineering, analysis of the impact of natural hazards on buildings, and computer technology for visualization and analysis in construction. A budget of $400,000-$500,000, half of which is allocated to building-related research, is supported by federal agencies, including the National Science Foundation, the National Park Service, and the U.S. Geological Survey (USGS), and by construction and computer companies.

DISTINCTIVE ATTRIBUTES: A broad perspective on architecture and hazard analysis, as well as a strong computer orientation.

PUBLICATIONS: Internal publications, as well as journals, conference proceedings.

UNIVERSITY OF ILLINOIS

School of Architecture
Champaign, IL 61820
217/333-1330

CONTACT: R. Alan Forrester, Director

MISSION AND FOCUS OF RESEARCH: Research is largely focused on energy use and the implications of energy-efficient techniques for building design; lighting; impact of artificial intelligence on building design; history and preservation; materials; and building types and residential developments, particularly user and behavioral aspects. Facilities include a lighting simulation laboratory as well as the full use of Civil Engineering facilities, the U.S. Army Corps of Engineering Research Laboratory (CERL), and the Architectural Research Center. There are 50 faculty members and graduate students. The research budget is approximately $200,000. Sources are grants from government agencies, corporations, local government, and overseas government support.

PUBLICATIONS: Journals, theses, conference proceedings.

ILLINOIS INSTITUTE OF TECHNOLOGY

Architecture Department
3360 South State Street
Chicago, IL 60616

CONTACT: George Schipporeit, Chairman

MISSION AND FOCUS OF RESEARCH: Research projects include prefabricated housing technology, "house of the future," robotics in family settings, and technology to provide an energy-on-demand capability for housing.

DISTINCTIVE ATTRIBUTES: Investigation of new construction technologies, high-rise construction techniques, application of high technology to offices and homes.

PUBLICATION: Internal newsletter.

UNIVERSITY OF MARYLAND

Architecture and Engineering Performance Center (AEPIC)
3907 Metzerott Road
College Park, MD 20742
301/935-5544
301/454-3428

CONTACT: John Loss, Professor of Architecture and Executive Director of AEPIC

MISSION AND FOCUS OF RESEARCH: AEPIC's work emphasizes the performance of buildings and constructed facilities. The center collects, classifies and analyzes data on problems, malfunctions, and failures in architectural and civil projects, their histories and outcomes. Opened in 1982 with a staff of seven, the center is funded by patron membership, subscription membership, and indirect assistance from research grants. More than half its annual effort will be devoted to the problems of buildings.

DISTINCTIVE ATTRIBUTES: The center is the first and only one of its kind.

PUBLICATIONS: Newsletter: *Architecture & Engineering Performance Notes*.

MASSACHUSETTS INSTITUTE OF TECHNOLOGY

Laboratory of Architecture and Planning
Building 4, Room 209
77 Massachusetts Ave.
Cambridge, MA 02139
617/253-1350

CONTACT: Michael Joroff, Director

MISSION AND FOCUS OF RESEARCH: The laboratory's research focuses on new materials, building performance, the impact of telecommunications on urban forms and building design, internationalization of planning practices in the building industry and design methodology. Wind tunnels, demonstration houses, lighting laboratories, CAD, and computer support are available. About $2 million a year is allocated to building-related research. Sources are government and industry, including foreign industry.

DISTINCTIVE ATTRIBUTES: Interdisciplinary, international perspective.

PUBLICATIONS: Internal publications.

UNIVERSITY OF MICHIGAN

Architecture and Planning Research Laboratory
College of Architecture and Urban Design
2000 Bonisteel Boulevard
Ann Arbor, MI 48109
313/764-1340

CONTACT: Colin W. Clipson, Director

MISSION AND FOCUS OF RESEARCH: The Research Laboratory's goal is to develop new knowledge and methodologies. There is an advisory service and consulting program for planning and design problems. Specific areas of research are environmental planning, building technology, facility and energy management, human behavior and the environment, computer-aided building design, building evaluation, policy planning, and building forms and land uses. The staff includes nine faculty members, four support staff, and graduate students. Support comes from the parent institution, U.S. government, foundations, and industry.

The Building Technology Laboratory has a 260-square-foot test room for thermal, luminous, and acoustical environments to be manipulated; solar and daylighting simulators; structural testing on a 30- x 24-foot testing floor; a hydraulic testing machine for stress/strain relationships; and a 5-ton overhead crane.

The Computer Laboratory uses the computer facilities of the university, which has a large virtual memory machine. This program develops computer-aided building software tools. Design application programs include thermal, lighting, structural, architectural, site, fire safety, and handicapped access analysis programs. It can make two- and three-dimensional building models.

The Facility and Environmental Simulation Laboratory has 2,200-square-foot of indoor space, and an outdoor area for building simulations. It is the site for the study of thermal, acoustical and lighting problems.

The Visual Studies Laboratory has 3,000 square feet of area including a darkroom and studio work areas.

PUBLICATIONS: Internal publication; conferences and workshops.

UNIVERSITY OF MINNESOTA

School of Architecture and Landscape Architecture
110 Architecture
89 Church Street
Minneapolis, MN 55455
612/624-0066

CONTACT: Harrison Fraker, Head

MISSION AND FOCUS OF RESEARCH: The focus of research is on building energy research with an emphasis on technology transfer, building energy conservation studies, daylighting, behavioral studies, urban design, history and theory, and CAD. Facilities include a sky simulator for model testing and light testing and 20 complete CAD stations. The research budget is $1 million with about a third allocated to building-related research.

DISTINCTIVE ATTRIBUTES: Regional Daylighting Center; Minnesota Cold Climate Building Research Center; Computer-Aided Architectural Design Center

PUBLICATIONS: Project reports; *Midgard*, a journal of history/theory.

STATE UNIVERSITY OF NEW YORK

School of Architecture and
  Environmental Design
3435 Main Street
Buffalo, NY 14214
716/831-3483

CONTACT: Robert G. Shibley, Chairman

MISSION AND FOCUS OF RESEARCH: SUNY Buffalo's research program focuses on computer-aided design and human factors of design including environmental behavior, handicap accessibility, design processes, and children's playgrounds. Facilities include an adaptive environment laboratory, a computer-aided design laboratory, and geometric solid modeling capability. The budget for computer-assisted design is between $350,000 and $400,000; and for human factors is between $200,000 and $350,000 depending on funding sources.

OHIO STATE UNIVERSITY	Department of Architecture
189 Brown Hall
190 West 17th
Columbus, OH 43210
614/422-5567

CONTACT: Robert Livesay, Chairman

MISSION AND FOCUS OF RESEARCH: Related research focuses on computer-aided architectural design to develop 3-D architectural models and from them to produce drawings and environmental simulation. There is a staff of five faculty members and graduate students. The department has a budget of $1.3 million. Recently, they received a $500,000 contract from IBM. The facilities include an environmental simulation laboratory that has a wind tunnel, daylight simulation, an acoustical tank, and a light box. The CAD laboratory is one of the most advanced. It has an IBM 4341 computer with virtual memory. Currently, it has 4 high-resolution vector refresh screens and is adding 12 additional workstations. It has a technologies graphics processor with the ability to generate 16 colors. Software is developed in-house.

PUBLICATIONS: Journals, conference proceedings, users' manuals for software, and theses.

UNIVERSITY OF OREGON	Department of Architecture
School of Architecture and Allied Arts
University of Oregon
Eugene, OR 97403
503/686-3656

CONTACT: Donald B. Corner, Head

MISSION AND FOCUS OF RESEARCH: Areas of research are directly related to faculty interests. Work includes environmental control systems, climate analysis and building design using CAD, solar energy, daylight, electric lighting, housing design and production, and flood and earthquake emergency housing strategies. There are 45 faculty members and a support staff of four or five. Facilities include computer facilities for environmental control research.

PRINCETON UNIVERSITY	School of Architecture
Princeton University
Princeton, NJ 08544
609/452-3729

CONTACT: Robert Gutman

MISSION AND FOCUS OF RESEARCH: The research program includes studies of the technology of historic structures, including engineering analysis of important historic buildings; the history of current trends in architectural practice; the changing nature of architectural forms; and the size of the profession.

DISTINCTIVE ATTRIBUTES: Computer modeling of structures and simulating stress in historic structures. The combinations of engineering and architectural history, and sociology and architectural history, are also distinctive.

RENSSELAER POLYTECHNIC INSTITUTE

Center for Architectural Research
Troy, NY 12181-3590
518/266-6461

CONTACT: Walter Kroner, Director

MISSION AND FOCUS OF RESEARCH: The center focuses on five major topics: the impact of space and space layouts, innovative building codes that improve safety, industrialized building systems, energy research, and emerging technologies in architecture. The center has a daylighting laboratory, artificial sky, and computer support. The budget is between $250,000 and $500,000, from DOE and state governments.

DISTINCTIVE ATTRIBUTES: The oldest department of architectural research in the United States, and sited within a technical institute.

PUBLICATIONS: Journals, conference proceedings.

THE UNIVERSITY OF TEXAS AT AUSTIN

Center for the Study of American Architecture
School of Architecture
Austin, TX 78712
512/471-1922

CONTACT: Lawrence W. Speck, Director

MISSION AND FOCUS OF RESEARCH: The center is an integral part of the School of Architecture at the University of Texas at Austin. It conducts thorough examinations and analyses of the history and development of architecture in the United States. There is a staff of three full-time faculty and research assistants. Support comes from endowments, industry, and trusts. The center has access to 25,000 drawings (many of which are technical drawings), documentations of building processes, and the architecture library and rare book collection.

PUBLICATIONS: Annual journal, sponsors exhibitions, annual spring symposium, and other publications and symposia.

VIRGINIA POLYTECHNIC INSTITUTE AND STATE UNIVERSITY

Washington-Alexandria Center
101 North Columbus Street
Alexandria, VA 22314
703/548-0099

CONTACT: Frederick Krimgold

MISSION AND FOCUS OF RESEARCH: Areas of building-related research at VPI's Washington-Alexandria Center and in Blacksburg, Virginia, include architecture, building materials, civil engineering, and forestry. Particular areas of focus include international building development and planning, building economics, building materials, building technology, energy and environmental control systems, illumination, fiber-reinforced plastics, performance, ventilation, construction management, geotechnical engineering, composite materials, polymers, wood preservation, wood adhesives, applied manufacturing, and tool design and evaluation.

Facilities at the Washington-Alexandria Center and in Blacksburg include the Environmental Systems Laboratory, which maintains a structural testing laboratory and a fiber-reinforced plastics facility; an artificial sky; a low-speed (0-15 mph) wind tunnel for scale models (4' x 8' cross section); the William H. Sardo Jr., Pallet and Container Research Laboratory, which has 7,200 square feet of floor space with testing equipment, environmental chambers, and computer terminals; and a CAD laboratory.

DISTINCTIVE ATTRIBUTES: Lighting dome, wind tunnel, and computer-aided design system.

PUBLICATIONS: Journals, laboratory bulletins, conference proceedings, and project reports.

UNIVERSITY OF WASHINGTON

Center for Planning and Design
College of Architecture and Urban Planning
206 Architecture Hall, AL-15
Seattle, WA 98105
206/545-0930

CONTACT: Judith H. Heerwagen, Associate Director

MISSION AND FOCUS OF RESEARCH: Much of the work is conducted in conjunction with the engineering department. Energy studies include thermal performance of houses built to existing state energy codes and proposed regional conservation plans; computer simulations of energy usage for passive solar and traditional buildings, including thermal comfort analysis, daylight analysis, and projections of cost-effectiveness and life-cycle costing; daylight studies of the relationships between fenestration design, heat gain/loss, daylight and electric illumination, laboratory analysis of discomfort glare, merger of electric and daylight illumination strategies; passive solar strategies; and occupant responses to energy strategies.

Facilities include an atmospheric radiation measuring station, a regional daylighting center with an overcast sky simulator, a direct beam sunlight simulator with sample lamps and lighting fixtures, a passive solar test cell, test houses and a weather station, and computer facilities. Recent funding has been at approximately $12 million a year. All funding comes from outside sources.

DISTINCTIVE ATTRIBUTES: Four test houses that automatically record thermal performance and occupant use of houses including door/window usage, thermostat settings, and appliance usage.

PUBLICATIONS: Journals and conference proceedings.

UNIVERSITY OF WISCONSIN-MILWAUKEE

Center for Architecture and Urban Planning Research
University of Wisconsin-Milwaukee
Milwaukee, WI 53201
414/963-4014

CONTACT: Gary T. Moore, Director

MISSION AND FOCUS OF RESEARCH: Activities include environment-behavior research, policy planning, transportation systems, urban design, computer-aided design, lighting, energy, and architectural theory.

Annual budget of $750,000 is supported by World Bank, U.S. Army Corp of Engineers, National Endowment for the Arts, Graham Foundation, Wisconsin Department of Transportation, federal and state government, private foundations.

PUBLICATIONS: 110 titles and independent reports.

Engineering

UNIVERSITY OF ALASKA

Institute of Northern Engineering
Duckering Building
306 Tanana Drive
Fairbanks, AK 99701
907/474-7775

CONTACT: Thomas D. Roberts, Director

MISSION AND FOCUS OF RESEARCH: The institute concentrates on practical engineering problems associated with northern environments. In building research, topics include foundation stabilization in frozen soil, and thermal performance (heat loss) of building components; $600,000 of a $3 million budget is allocated to building research. The Department of Energy, the State of Alaska, and the U.S. military agencies are sources.

DISTINCTIVE ATTRIBUTES: Cold regions engineering research.

PUBLICATIONS: Technical journals.

UNIVERSITY OF CALIFORNIA

Earthquake Engineering Research Center
1301 South 46th Street
Richmond, CA 94804
415/231-9509

CONTACT: Patrick Quinn, Laboratory Manager

MISSION AND FOCUS OF RESEARCH: The center analyzes the dynamic response of buildings to earthquakes in an attempt to predict more accurately how these structures will respond and to prevent loss of life or property damage. This includes analysis of the architectural elements, including lights and windows. The characteristics and intensities of earthquakes and their effects on structural, mechanical, and soil systems are studies for purposes of prediction. The center also studies seismic effects on earth, soil-structure interaction, and seismic-resistant devices.

Facilities include a 20-square-foot shake table, other types of dynamic structural test facilities, a 50- x 30-foot structural test bed, extensive static and pseudodynamic facilities to evaluate earthquake resistance of typical building construction, and a 4-million-pound press. They also have access to a library of computer programs. There are 25-30 research professionals, some of whom are faculty members, and graduate students. Support comes from the U.S. government, the State of California, and other sources.

DISTINCTIVE ATTRIBUTES: Twenty-square-foot shake table with 140,000-pound capability.

PUBLICATIONS: Project reports, *EERC News* (a newsletter), and *Journal in Earthquake Engineering*.

UNIVERSITY OF CALIFORNIA

Structural Engineering Materials Laboratory
Davis Hall
Berkeley, CA 94720
415/642-3464

CONTACT: Roy Stephen

MISSION AND FOCUS OF RESEARCH: The Laboratory studies the properties of structural materials, including heavy concrete, composite materials, and steel; structural responses to earthquakes; structural analysis and design; and the mechanics of complex structural elements and systems. There is a research staff of 21 faculty members and 150 graduate students. Support for the $3 million budget comes from the parent institution, the U.S. government, and industry. Facilities include a fire research facility and test facilities including a 4-million-pound universal testing machine, and a large tie-down test floor. The Lab is associated with the Earthquake Engineering Research Center.

PUBLICATIONS: Project reports and journals.

CLEMSON UNIVERSITY

Department of Civil Engineering
Clemson, SC 29633
803/656-3314

CONTACT: R. H. Brown

MISSION AND FOCUS OF RESEARCH: The department focuses on structural analysis of masonry walls, wind effects on metal and masonry buildings, and building code changes that may be necessary to mitigate earthquake damage. There are 17 full-time faculty. Facilities include structural, concrete, hydraulic modeling and soil mechanics laboratories. The budget ranges from $500,000-$800,000, of which 30-40 percent is allocated to building research. The National Science Foundation, the Department of Energy, Sea Grant, State Highway Department, and private industry are sources.

PUBLICATIONS: Reports as required by contract, dissertations.

COLORADO STATE UNIVERSITY

Department of Civil Engineering
Fort Collins, CO 80523
303/491-8557

CONTACT: James Goodman, Professor

MISSION AND FOCUS OF RESEARCH: Areas of work include wood material properties and wood use in construction; structural design and analysis; large-scale glue-laminated structures; steel; concrete; reliability design procedures and their application in trusses; and floors, walls, and roofs. The facilities include a wood science lab, a flexible structural lab with tie-downs in the floor, equipment for breaking wood poles, testing machines, associated data-logging facilities, saws, and drawing facilities. The research budget is $500,000, of which $100,000 is used for building-related work, supplied by the the National Science Foundation and industry.

DISTINCTIVE ATTRIBUTES: The wood engineering program is a unique combination of wood science and technology with civil engineering.

PUBLICATIONS: *Structural Research Reports*; journals, reports, and papers.

COLORADO STATE UNIVERSITY

Solar Energy Applications Laboratory
College of Engineering
Fort Collins, CO 80523
303/491-8617

CONTACT: Tom Brisbane, Research Associate

MISSION AND FOCUS OF RESEARCH: The laboratory focuses on the integrated research and development of heating, cooling, and storage systems in solar residential systems. The program includes complete acquisition and monitoring to integrate all components of residential work. There is a full-time senior staff of 10. Facilities include three residential-type buildings. The research budget of under $1 million comes from the Department of Energy.

PUBLICATIONS: Reports, papers, and journals.

UNIVERSITY OF FLORIDA

Solar Energy and Energy Conversion Laboratory
Department of Mechanical Engineering
Gainesville, FL 32611
904/392-0820

CONTACT: Erich A. Farber, Director

MISSION AND FOCUS OF RESEARCH: The laboratory conducts studies in the absorption, transmission, and conservation of energy. Major topics are temperature and humidity control, the economics of solar energy, solar heating and cooling, and the application of solar technology to a variety of energy needs. Facilities include solar energy and energy conversion laboratories, test houses including two mobile homes, a concrete block building, and a frame building. About 25-30 percent of a budget that ranges from $250,000 to $1 million is allocated to building research. Funding sources are state governments, the Department of Energy, the National Bureau of Standards, the Department of Housing and Urban Development, and industry.

DISTINCTIVE ATTRIBUTES: Experience in the field and the availability of a hot, humid climate for testing.

PUBLICATIONS: State government channels; workbooks and pamphlets for public use.

GEORGIA INSTITUTE OF TECHNOLOGY

Center for Rehabilitation Technology
College of Architecture
Atlanta, GA 30332
404/894-4960

CONTACT: Richard L. Martin, Director

MISSION AND FOCUS OF RESEARCH: The center performs basic and applied research in response to the needs of the disabled while emphasizing the importance of incorporating barrier-free design concepts into the building design process. Facilities include a computer laboratory, a construction laboratory, and the use of all laboratories at the university, including the civil engineering laboratory. There is a core staff of 12 people. Yearly overhead funding for the staff is about $600,000 and is supplied by the state. This budget is augmented by projects that total about $1 million, supplied primarily by federal agencies.

PUBLICATIONS: Project reports and university publications.

UNIVERSITY OF ILLINOIS

Ceramic Engineering Department
204 Ceramics Building
105 South Goodwin
Urbana, IL 61801
217/333-3125

CONTACT: C. G. Bergeron
Richard Berger

MISSION AND FOCUS OF RESEARCH: Various members of the faculty focus on topics involving materials research on cement and concrete, gypsum board, and gypsum plaster. These include investigation into corrosion resistance, coatings, cement, refractory material directed at strength, wear, weather testing, dynamic loading of concrete, and fiber reinforcement. The department's budget is between $1.5 and $2 million, of which less than 10 percent is allocated to building research. The U.S. Army Corps of Engineers is a source of funding.

DISTINCTIVE ATTRIBUTES: Doing basic research in cementitious systems. The laboratory facilities include mercury pure symmeters, thermal evolved gas, an extensive concrete laboratory, and environmental chambers.

PUBLICATIONS: Professional journals, university publications.

UNIVERSITY OF ILLINOIS

Civil Engineering Department
1106 Newmark Civil Engineering Laboratory
Urbana, IL 61801
217/333-6948

CONTACT: H. Walker, Associate Department Head
W. J. Hall, Department Head

MISSION AND FOCUS OF RESEARCH: The department focuses its research on structural engineering (steel and concrete), particularly with respect to structural integrity and earthquake resistance; material science (cement); construction management; solar mechanics and foundations; photogrammetry; environmental engineering; hydraulics and water resources. Work in computer applications is extensive. The budget is approximately $3 million, all of which is allocated to building research. There are 78 professional staff, 40 support staff, and 120 research assistants. The facilities include a full-scale, two-story masonry building, structural testing laboratory, and computer support.

DISTINCTIVE ATTRIBUTES: One of the largest structural test facilities in the United States began research in 1903 and has a long history in the field. Many of the faculty serve on code-writing committees for steel, transportation, and materials.

PUBLICATIONS: Internal series of research reports and professional journals.

UNIVERSITY OF ILLINOIS

Small Homes Council-Building Research Council
1 East Street
Champaign, IL 61820
217/333-1910

CONTACT: Donald Percival, Research Professor of Wood Technology

MISSION AND FOCUS OF RESEARCH: The council's emphasis is on structural components for light-frame construction, primarily floor and roof trusses of wood. Its mission is to serve as an information source on building materials, directed at both contractors and the public.

IOWA STATE UNIVERSITY

Department of Mechanical Engineering
Ames, IA 50011
515/294-6886

CONTACT: Ron Nelson

MISSION AND FOCUS OF RESEARCH: Major research projects focus on heating, ventilation, air conditioning, and indoor air quality, using an energy research house, two environmental control chambers, and an airflow loop.

PUBLICATIONS: ASHRAE reports and conference proceedings.

IOWA STATE UNIVERSITY

Structural Research Laboratory
Ames, IA 50011
515/294-7456

CONTACT: W. W. Sanders, Jr., Associate Director

MISSION AND FOCUS OF RESEARCH: Research related to structural engineering is performed, including studies of prestressed concrete, reinforced concrete, welding, steel structures, composite cold-formed slabs, materials fatigue, and behavior of structural members and of structural connections. Facilities include a 1-million- pound capacity structural test floor, 400,000-pound universal and 110,000-pound Falgram closed-loop test machines, machine shops, and a multipurpose laboratory. There is a staff of nine research professionals, two technicians, and four supporting professionals.

PUBLICATIONS: Journals.

LEHIGH UNIVERSITY

Center on Advanced Technology
for Large Structural Systems (ATLSS)
Bethlehem, PA 18015
215/758-3515

CONTACT: John W. Fisher, Professor and Director

MISSION AND FOCUS OF RESEARCH: The ATLSS center is a focal point for research that will lead to technological developments to benefit structures-related industries in design, fabrication, construction, and inspection and protection. The research areas emphasized are:

o Development of advanced design concepts and advanced connections methods for large structural systems (buildings, bridges, cranes, platforms, vehicles, vessels, etc.) and integration of these concepts and methods into fabrication and construction.
o Development of new and improved methods of in-service monitoring and protection.
o Development of computer-based expert systems to facilitate collaborative decision making among designers, fabricators, and inspectors.
o The quantitative economic issues of new or proposed systems.

Facilities include a multidirectional testing facility for three-dimensional loading tests on full-size structural components. The floor of the facility is 100 x 40 feet and there are right-angled vertical reaction walls with heights ranging from 20 to 50 feet so that two-direction lateral loads, as well as vertical loads, can be applied. The facility has computer-controlled loading and high-speed computer-based data acquisition. The center has an annual research budget of $2 million, from private and government sources.

DISTINCTIVE ATTRIBUTES: The multidirectional testing facility is unique in the United States.

PUBLICATIONS: ATLSS reports, quarterly newsletter, journal papers, and conference proceedings.

LEHIGH UNIVERSITY

Fritz Engineering Laboratory
Department of Civil Engineering
Building 13
Bethlehem, PA 18015
215/861-3531

CONTACT: Lynn S. Beedle, Director

MISSION AND FOCUS OF RESEARCH: The laboratory's efforts are generally focused on civil engineering and related disciplines with research programs and industrial testing facilities. Studies are done in related structural fields, including structural steel, building systems, fatigue and fracture, structural concrete, structural connections, and structural stability. Other project areas include hydraulics, geotechnical engineering, and environmental engineering.

Facilities include a two-story and a seven-story unit for testing large structural members. The research is interdisciplinary, involving other centers and departments, such as the Materials Research Center and the Center for Surface Coating Research. Laboratories within the Fritz Laboratory include concrete, hydraulics, instruments, materials testing, sanitary engineering, geotechnical engineering, structures, and welding. Equipment includes strain measurement devices, automatic recording equipment, high-speed cameras, structural model instruments, and geotechnical field test equipment. An active research staff includes 20 faculty members and about 40 graduate students with a budget of nearly $1 million.

DISTINCTIVE ATTRIBUTES: 70 x 130 foot, seven-story structure and a 5-million-pound testing machine.

PUBLICATIONS: Journals.

LEHIGH UNIVERSITY

Institute for the Study of High-Rise Habitats
Building 13
Bethlehem, PA 18015
215/758-3515

CONTACT: Lynn S. Beedle

MISSION AND FOCUS OF RESEARCH: The institute's research and educational focus is on studies of the technological and socioeconomic aspects of tall buildings. The institute is a part of the Council on Tall Buildings and Urban Habitat. Work is done on the structural, mechanical, and architectural aspects of tall buildings, the livability of projects, their appropriateness to the context for which they are planned, and their function as part of the urban design.

Specific areas of research include planning and design of tall buildings, earthquake resistance of high-rise building systems, performance evaluation of tall buildings under natural hazard environments, frame stability, urban services, seismic safety of prefabricated concrete buildings, and modeling human errors in structural design and construction. The institute also analyzes the impact of tall buildings on the local environment including the second century of the skyscraper.

Facilities include a knowledge-based CAD system. Offices are located in the Fritz Engineering Laboratory which gives the Institute access to the lab's facilities. About six faculty members are actively involved in research. The research budget is $250,000-$300,000 which comes from the National Science Foundation, Federal Emergency Management Administration, and other sources.

PUBLICATIONS: Journals and monographs on tall buildings; a series of books.

UNIVERSITY OF MARYLAND

Construction Engineering and Management Program
College of Civil Engineering
College Park, MD 20742
301/454-2438

CONTACT: Leonard E. Bernold, Assistant Professor

MISSION AND FOCUS OF RESEARCH: The program examines and analyzes the kinds of construction that are candidates for automation and robots. It also addresses the use of emulation and simulation for the design of automated flexible construction systems. The program does experimentation with computer-integrated construction processes (CAD/CAM) and also breaks down the time and cost of work on structures. The program has a $500,000 budget with three faculty members and 40 graduate students. Facilities include a minicomputer and microcomputer lab, construction robotics lab, video tape, and time- lapse equipment.

DISTINCTIVE ATTRIBUTES: The program is one of only two programs to offer expert simulation of construction activities. It has a close working relationship with the Mechanical Department and the National Bureau of Standards.

PUBLICATIONS: Internal reports, reports for sponsors as necessary, and professional journals.

MASSACHUSETTS INSTITUTE OF TECHNOLOGY

Center for Construction Research and Education
Department of Civil Engineering
Room 1-175
Cambridge, MA 02139
617/253-7273

CONTACT: Charles H. Helliwell, Deputy Director

MISSION AND FOCUS OF RESEARCH: The center contains all research and educational activities related to construction within the Civil Engineering Department. The principal areas of focus are management; financial, labor, and equipment resources; and technology. The technology section comprises computer applications in design and construction (expert systems, artificial intelligence, and CAD); automation and robotics (including automated conditions assessment systems for defining programs for maintenance and repair, and the development of robotic equipment for site construction tasks such as wall building, scaffolding, etc.); and engineered materials for repair, maintenance, and new construction.

Facilities include computer laboratories; a materials testing laboratory; a structural testing laboratory for shear, strain, and strength tests; a new robotics laboratory; and a facility for non-destructive evaluation of materials and structures in situ.

The departmental research budget is about $6.5 million, with about $2.3-$2.7 million for construction research. Funding comes from the National Science Foundation, U.S. Army Corps of Engineers, the State Department, U.S. AID, and industry. In September of 1986, the Center received a $15 million, 5-year grant from the DOD University Research Initiative to establish the Program for Advanced Construction Technology (PACT). The Center has 4-5 full-time research associates, 15-18 faculty members doing research, 15 graduate research assistants, and 10 PACT Fellowship students.

PUBLICATIONS: A newsletter two or three times a year, a publication series, and a department research publication series.

UNIVERSITY OF MINNESOTA　　Underground Space Center
790 Civil and Mineral Engineering Building
500 Pillsbury Drive, S.E.
Minneapolis, MN 55455
612/624-0066

CONTACT:　　Raymond L. Sterling, Director

MISSION AND FOCUS OF RESEARCH: Major research areas include building foundations, the uses of underground construction, earth-contact heat transfer, energy performance monitoring, geotechnical engineering, and underground urban planning. There are seven technical professionals on staff. Monitoring projects takes place at field experiment sites. In addition, a building foundation test facility is under development. The $300,000 research budget is provided by federal and state government, as well as private industry.

PUBLICATIONS: The center publishes its own journal, books, technical papers, and reports. It also sponsors conferences.

UNIVERSITY OF NEW MEXICO　　New Mexico Engineering Research Institute
P.O. Box 25
Albuquerque, NM 87131
505/844-5189

CONTACT:　　Delmar Calhoun, Director

MISSION AND FOCUS OF RESEARCH: The institute conducts research and development in the following areas: structural engineering, with emphasis on blast and shock effects; explosives effects, with emphasis on simulation of blast and shock environments; fire suppression; construction materials properties; pavements; soil mechanics; environmental science and engineering; instrumentation development; geographic information systems; and engineering computational analysis. The annual budget is approximately $20 million, two-thirds of which is for general building research. NMERI undertakes research for the U.S. Air Force, as well as other federal, state, local, and private clients.

DISTINCTIVE ATTRIBUTES: Large-scale testing of structures and components, high-explosives capabilities, multichannel dynamic data acquisition, and large-scale fire suppression tests.

PUBLICATIONS: Research results are generally published by sponsoring agencies.

STATE UNIVERSITY OF NEW YORK

College of Environmental Science and Forestry
State University of New York
Syracuse, NY 13210
315/470-6880

CONTACT: Leonard Smith

MISSION AND FOCUS OF RESEARCH: Major research areas include testing and research of laminated lumber, metal truss plates in roofs, arching of wood trusses in residential housing, wood preservative treatments, manufactured plywood and particle board. Facilities include a universal stress testing machine with 400,000-pound capacity; impact machine, conditioning chambers; finishing spray booth; wood working facilities; and a 32-foot truss tester.

DISTINCTIVE ATTRIBUTES: One of the best-equipped labs for wood research in an academic institution; roof truss testing machine.

PUBLICATIONS: Professional journals, university publications, and books.

NORTHWESTERN UNIVERSITY

Civil Engineering Department
2145 Sheridan Road
Evanston, IL 60201
312/491-3258

CONTACT: Raymond Krizek, Chairman

MISSION AND FOCUS OF RESEARCH: The research focuses structural engineering, foundations, concrete, steel, CAD, geotechnical engineering, and nondestructive testing. About $1.7 million is allocated to building-related research with funds coming from the federal government and private industry.

DISTINCTIVE ATTRIBUTES: The facility has a testing machine with triaxial and torsional test chambers, an axial load capacity of 1,100,000 pounds, fluid pressure to 20,000 pounds per square inch, and torque of 100,000 lb. in. Its temperature range is from room temperature to $600^\circ C$, and it can be pressurized with nitrogen, air, or water. Specimens can be either sealed or unsealed. The test cavity measures 8.5 inches; force and deformation can be measured inside the cavity. Loading is servocontrolled and has the potential for being computer controlled. It is a stiff machine, suitable for concrete and rocks, and can also be used for other materials.

Additional equipment includes a small electromagnetic shake table; a very-high-pressure (to 500,000 pounds per square inch) triaxial test machine for small specimens which can go from room temperature to $2000^\circ C$; small triaxial loading devices; soil testing equipment; water jet rock-cutting equipment; and fracture testing equipment.

PUBLICATIONS: Reports as required by contractors; journals.

OREGON STATE UNIVERSITY

Department of Construction Engineering Management
Corvallis, OR 97331
503/754-2006

CONTACT: Harold Pritchett

MISSION AND FOCUS OF RESEARCH: Management strategies in construction is the major research area at Oregon State. Related research includes concrete for walls, glue-laminated beams, and wood. Facilities include large presses and equipment for designing mixes to test for strength and wear. About $100,000 of the budget is used for building research, with private industry as the primary source.

PENNSYLVANIA STATE UNIVERSITY

Department of Architectural Engineering
Engineering Building A-Room 202A
University Park, PA 16802
814/865-8394

CONTACT: Stan Mumma

MISSION AND FOCUS OF RESEARCH: The department's research activities focus on acoustical studies, environmental work in indoor air quality and intelligent buildings, and structural research, as well as studies of lighting, heating, ventilating, air conditioning, building thermal response characteristics, energy storage, and building automation and control. There are 33 faculty members. Facilities include an illumination and acoustics laboratory, a building thermal response laboratory, a model testing laboratory, a CAD laboratory, and use of the structures and model testing laboratories with the Department of Civil Engineering. Funding is provided by state and federal agencies and by private industry.

PUBLICATIONS: Journals.

PENNSYLVANIA STATE UNIVERSITY

Engineering Research Program
101 Hammond Building
University Park, PA 16802
814/865-1804

CONTACT: Tom Seliga, Associate Dean for Graduate Studies and Research

MISSION AND FOCUS OF RESEARCH: The building-related work includes structural design and analysis, system analysis, air-conditioning ducting, illumination, energy consumption, heating systems, acoustics, and vibration analysis. Facilities include instrumentation for indoor environmental measurements, an acoustics laboratory, and a computerized design laboratory. About 10-15 faculty work in building- and construction-related areas, with a research budget exceeding $500,000.

PUBLICATIONS: Journals and reports.

PENNSYLVANIA STATE UNIVERSITY

Structures Laboratory
Civil Engineering Department
212 Sackett Building
University Park, PA 16802
814/865-8391

CONTACT: Harry West, Professor, Civil Engineering

MISSION AND FOCUS OF RESEARCH: While the bulk of the research is done on bridges, much of this can be applied to buildings. Work is also done on prestressed concrete, structural analysis, and the use of computers for theoretical studies and analytical design studies. There are four researchers in Civil Engineering. Facilities include computers, a loading frame with the capacity to test structural models and full-scale structures, and equipment for the racking of frames. The research budget is $1.5 million, with $300,000 allocated to structures-related research. The Pennsylvania Department of Transportation, FHA, and private industry are sources.

PUBLICATIONS: Reports are supplied to sponsors, journals.

PURDUE UNIVERSITY

School of Technology
Department of Building Construction and Contracting
Room 453, Knoy Hall
West Lafayette, IN 47907
317/494-2467

CONTACT: Steven Easley

MISSION AND FOCUS OF RESEARCH: Department objectives are to develop systems using new materials and techniques for the construction industry; to research applications of construction equipment, tools, and building methods to increase productivity and reduce construction costs; and to analyze cost-effective applications of new and existing building technology. Research activities are carried out in the areas of structural design; the study of materials, including steel, wood, concrete, and plastics; and testing of product applicability and use in construction. Some research is also performed in the development of cost-effective energy-efficient construction techniques to reduce heating and cooling costs in buildings and in the construction of superinsulated houses. There are 20 faculty members doing related research. Facilities include a construction laboratory with an overhead tower crane to allow for building within the laboratory to test materials and products; an observation deck for viewing and video taping in the laboratory; a materials laboratory; a soils laboratory; a CAD laboratory; and a thermal test booth.

DISTINCTIVE ATTRIBUTES: It is the second largest school of construction management in the United States. Undergraduate enrollment is approximately 400 students.

PUBLICATIONS: Professional journals and general media.

STANFORD UNIVERSITY

Foundation Engineering
Department of Civil Engineering
Stanford, CA 94315
415/723-0236

CONTACT: Raymond Seed, Assistant Professor

MISSION AND FOCUS OF RESEARCH: The program's emphasis is on soil analysis, including soil compaction and evaluation of compaction-induced stresses on structural performance. This is relevant to basement walls, bridges, and buried structures. The geotechnical laboratory has the capacity to perform soil-stress path and strain path tests, controlled triaxial testing, torsional shear testing, resonant column testing, and cyclic triaxial testing. Facilities include a shake table, a Ko-odemeter, and micro- and minicomputers. Two geotechnical faculty members and 12 graduate students do research. A budget of up to $125,000 is provided by government grants and industrial contributions.

PUBLICATIONS: Technical journals and conference proceedings.

STANFORD UNIVERSITY

John A. Blume Earthquake Engineering Research Center
Stanford, CA 94315
415/723-4129

CONTACT: Helmut Krawinkler, Co-Director

MISSION AND FOCUS OF RESEARCH: The center's efforts are focused on the design of buildings, improving seismic risk and hazard evaluation, and small-scale testing of building dynamics. There is a structures laboratory with two shake tables, one horizontal and one vertical; two MTS universal testing machines; a static test bed; two loading frames; and VAC minicomputer facilities. The research budget is about $500,000 and comes from the National Science Foundation, the U.S. Geological Survey, the Federal Emergency Management

Administration, the Electrical Power Research Institute, and private industry. The staff is composed of 9 faculty members, 6 research associates, and 30-40 graduate students.

DISTINCTIVE ATTRIBUTES: Unique in seismic risk and hazard evaluation and emphasis on small-scale testing of building dynamics.

PUBLICATIONS: Technical journals and report series.

STEVENS INSTITUTE OF TECHNOLOGY

Building Technology Center
Department of Civil Engineering
Hoboken, NJ 07030
201/420-5360

CONTACT: Thomas P. Konen, Director

MISSION AND FOCUS OF RESEARCH: The center conducts related research in the areas of water supply and drainage systems for buildings, heat transfer through windows and other envelope penetrations, product development for the plumbing industry, piping system analysis, passive solar systems, and fire safety systems. Facilities include a 10-story plumbing tower, computer facilities, and use of the mechanical engineering laboratories. The research budget is $200,000 to $300,000, and there is a staff of 12.

DISTINCTIVE ATTRIBUTES: The 10-story water supply and drainage tower; fire protection studies.

PUBLICATIONS: Journals, laboratory reports, and conference proceedings.

THE UNIVERSITY OF TEXAS AT AUSTIN

Architectural Engineering Group
ECJ 5.208
Austin, TX 78712
512/471-1732

CONTACT: David W. Fowler, Director

MISSION AND FOCUS OF RESEARCH: The group's emphasis is on testing materials and complete buildings, regarding their rehabilitation, durability, construction, and environmental systems. Research on masonry includes concrete, steel, and wood. Facilities include a large testing lab with a reaction wall for three-story buildings, a large testing floor for vertical and hydraulic testing, a data acquisition system, and environmental chambers. Activities include: earthquake simulation; CAD work; testing for abrasion and strength; durability; water permeability; freeze and thaw effects; and elasticity of basic connection materials. The budget for structural research, including civil engineering, is $3 million from federal and state agencies, with about one-half allocated to building-related work. There are 20 faculty members, 125 graduate students, and 10-15 technical staff.

PUBLICATIONS: Journals, reports, and conference proceedings.

THE UNIVERSITY OF TEXAS AT AUSTIN

Construction Industry Institute
3208 Red River, Suite 300
Austin, TX 78705
512/471-4640

CONTACT: Richard L. Tucker, Director

MISSION AND FOCUS OF RESEARCH: CII is a consortium of 57 companies and 15 universities. It is a research institute created to develop and disseminate information to advance the cost effectiveness of the construction industry. Major areas of study include full-scale modeling in reinforced concrete, lateral loading, seismic studies, structural framework, steel structural research with manufactured houses, acoustical properties, geotechnical work for foundations, and masonry and materials. The CII uses all laboratories and facilities of the University of Texas including the Ferguson Structural Engineering Laboratory, Center for Polymer Research, Geotechnical Engineering Laboratory, and climate control room and has use of the facilities of other member universities. The research budget is about $1 million, which comes primarily from member companies that pay $25,000 a year each in the form of an unrestricted grant. CII operates with about 50 faculty members and 50 staff members, and also uses faculty members and graduate students from participating universities.

PUBLICATIONS: Issues its own reports and journals.

THE UNIVERSITY OF TEXAS AT AUSTIN

Geotechnical Engineering Center
Department of Civil Engineering
ECJ 6.2
Austin, TX 78712
512/471-1555

CONTACT: Lymon Reese, Director

MISSION AND FOCUS OF RESEARCH: The center identifies soils susceptible to damage during earthquakes, tests the dynamic properties used for analysis, works on pile foundations and drilled shafts, evaluates the behavior of piles under lateral loads, and studies rock mechanics and tunneling. The research staff includes 6 faculty members and 50 students. The center has several soil mechanics laboratories, field study facilities, and equipment for on-site field tests. The research budget is about $500,000 from industry and contracts.

PUBLICATIONS: Papers, reports, and journals.

THE UNIVERSITY OF TEXAS AT AUSTIN

Phil M. Ferguson Structural Engineering Laboratory
Balcones Research Center
10100 Brunet Road, Building 24
Austin, TX 78758
512/471-7259

CONTACT: Karl H. Frank, Secretary, Research Council on Structural Connections

MISSION AND FOCUS OF RESEARCH: The Ferguson Laboratory designs, constructs, analyzes, and tests structural steel, reinforced concrete, and prestressed concrete. It also does work on masonry construction and wood-frame construction.

The laboratory is a 50,000-square-foot facility with 30,000 feet of tie-down floors. It maintains loading facilities for fatigue tests and has 600 channels of data acquisition, a variety of closed loop test equipment for fatigue and earthquake testing, and vertical tie-down walls. The operating staff includes 8 faculty members, 12 technical staff, and 50 students. They work with a budget of $1 to $1.2 million, contributed equally by industrial organizations, the state highway department, and the federal government. Approximately 30 percent of the budget is allocated to building-related research.

DISTINCTIVE ATTRIBUTES: Has the largest structural engineeering laboratory related to a university.

PUBLICATIONS: Technical journals.

UNIVERSITY OF WASHINGTON

Department of Civil Engineering
Seattle, WA 98195
206/543-2390

CONTACT: Neil Hawkins

MISSION AND FOCUS OF RESEARCH: The department focuses on structural engineering and design, construction, water supply, surveying, hydraulics, geotechnical engineering, and seismic studies.

DISTINCTIVE ATTRIBUTES: There are eight laboratories, including computer, soils, construction, and structural laboratories. The structural laboratory is 6,000 square feet and has a 2.2-million-pound testing machine with 20-foot pulling distance. There are also a wide variety of servocontrolled jacks, servocontrolled testing machines, and automatic data acquisition and reduction equipment. There is a 6-foot x 6-foot shake table that is 5 feet deep and movable with a retaining wall at one end.

PUBLICATIONS: Independent reports, some in journals.

UNIVERSITY OF WISCONSIN-MADISON

Engineering Experiment Station
Engineering Research Building
1500 Johnson Drive
Madison, WI 53706
608/263-1601

CONTACT: Clayton Smith, Assistant Dean

MISSION AND FOCUS OF RESEARCH: Related work focuses on structures design; extensive work with materials and their strength, including steel, concrete, wood, and composites; and designing solar systems for heating and cooling. Facilities include computers for modeling and facilities for structures and materials testing. The college has a staff of 200 faculty members, 1,000 graduate students, and 300 supporting staff. They work with a $26 million research budget, an increasing percentage of which is allocated to building-related research. Budget sources are 60 percent federal government, 25-30 percent industry, and the rest from foundations and internal sources.

PUBLICATIONS: Journals and reports.

FEDERAL LABORATORY PROFILES

THE ACID DEPOSITION RESEARCH PROGRAM

U.S. Environmental Protection Agency
RD 680
401 M Street, SW
Washington, DC 20460

CONTACT: Barbara Levinson, Program Manager

MISSION AND FOCUS OF RESEARCH: The Acid Deposition Research Provision (ADRP) is part of the National Acid Rain Precipitation Assessment Program, the members of which include the National Park Service, the Bureau of Mines, and the Environmental Protection Agency (EPA). The EPA uses both laboratory and field exposures to study the relationship between acid rain and materials damage. Special attention is focused on the degradation of paint and metals. Work is also being done on estimating how many materials are at risk from acid rain damage, and the potential costs. Research findings may include information on weathering of certain materials and the means to measure paint damage over time. EPA'S 1987 budget for this program is $2.7 million. Most of the damage function research is being done at EPA's Atmospheric Sciences Laboratory in

Research Triangle Park, North Carolina. The inventory of painted surfaces is being done through EPA's Environmental Monitoring Lab in Las Vegas, Nevada.

ACID RAIN RESEARCH PROGRAM

National Park Service
P.O. Box 37127
Washington, DC 20013
202/343-1055

CONTACT: Susan Sherwood, Cultural Resources, Acid Rain

MISSION AND FOCUS OF RESEARCH: The program studies the effects of air pollution on building stone and bronze, especially in historic buildings and statues. Based on a budget of $800,000 per year for 10 years, research is contracted to federal and academic laboratories. The program is part of the National Acid Rain Precipitation Assessment Program.

AIR AND ENERGY ENGINEERING RESEARCH LABORATORY

ENVIRONMENTAL MONITORING SYSTEMS LABORATORY

HEALTH EFFECTS RESEARCH LABORATORY

U.S. Environmental Protection Agency
Research Triangle Park, NC 27711
919/541-2821

CONTACT: W. Gene Tucker, Chief, Indoor Air Branch, Combustion and Indoor Air Division, Air and Energy Engineering Research Laboratory

MISSION AND FOCUS OF RESEARCH: The three laboratories at Research Triangle Park work together on the EPA's Indoor Air Quality Program. Major research areas include development of methods to measure indoor air quality; testing of emissions from indoor sources such as building materials, furnishings, and combustion devices; development and testing of indoor air pollutant control methods, especially for radon; research on the health effects of indoor air pollutant mixtures; and monitoring of indoor air quality in buildings to estimate indoor air pollutant exposures. Research results are published in the research literature and in public information documents. On-site facilities are used for some research, but the majority is done through contractual arrangements with universities and R&D organizations. An annual budget of approximately $2 million is devoted to indoor air quality R&D. An additional $1.5 million is applied to the development and testing of radon reduction techniques for homes.

PUBLICATIONS: Technical journals and EPA publications.

BROOKHAVEN NATIONAL LABORATORY
(Polymer-Concrete Development Laboratory)

U.S. Department of Energy
Department of Applied Sciences
Building 526
Upton, NY 11973
516/282-3036

CONTACT: Meyer Steinberg, Head of Process Sciences Division

MISSION AND FOCUS OF RESEARCH: Brookhaven maintains a test facility to perform basic and applied research on materials that are composites of polymer and aggregate. The lab has been responsible for the development of polymer concrete and pipe coatings and aggregates bound with resin. Materials are primarily used in pipes and wells to resist acids, alkalis, and chemicals.

BROOKHAVEN NATIONAL LABORATORY
(Test House and Space Conditioning Equipment Laboratories)

Department of Applied Sciences
Building 120
Upton, NY 11973
516/282-7726

CONTACT: John Andrews, Head, Architectural and Building Systems Group

MISSION AND FOCUS OF RESEARCH: Brookhaven has developed a case study approach to testing and monitoring alternative building methods for energy efficiency. The lab continuously monitors test houses and conducts energy analysis using a variety of materials and methods of construction, including an international housing village to test the efficiency of building methods of other countries. Brookhaven also manages a Heat Pump Laboratory, which can be used for transient or steady-state tests of liquid-source heat pumps or of individual heat pump components, and a Combustion Equipment Technological Laboratory to measure performance and thermal efficiency of oil- and gas-fired furnaces and boilers. The staff at this facility numbers 20 with a budget of $2 million.

PUBLICATIONS: Lab reports, professional journals, and conference publications.

CENTER FOR BUILDING TECHNOLOGY

National Institute of Standards and Technology
Building 225
Gaithersburg, MD 20899
301/975-5900

CONTACT: Richard N. Wright, Director
Charles Culver, Chief, Structures Division
James E. Hill, Chief, Building Environment Division
Geoffrey Frohnsdorff, Chief, Building Materials Division

MISSION AND FOCUS OF RESEARCH: The center's three divisions--Structures, Building Environment, and Building Materials--perform analytical, laboratory, and field research in areas of engineering and science pertinent to the usefulness, safety, and economy of buildings. The center also develops technology to predict, measure, and test the performance of building materials, components, and practices. Descriptions of each division's facilities follow.

The budget for 1987 is about $12 million. One-third of this is directly appropriated by the Congress while the remainder comes from other federal agencies. The center has a staff of about 150, of which 100 are professionals, half of whom hold doctorates.

Structures Division This division seeks to increase the productivity and safety of building construction by providing the basis for improved structural and earthquake criteria. Its laboratory includes a large-scale structural testing facility with a 12-million-pound universal testing machine, capable of simulating axial and lateral loads simultaneously on large-scale components up to 60-feet tall; a computer-controlled tri-directional structural testing facility, capable of applying loads simultaneously in three directions, which can study earthquake and wind effects; and a test floor on which beams, slabs, frames, or complete structures can be submitted to static loads or cyclic loading up to 50,000 pounds.

Building Environment Division This division attempts to reduce the cost of designing and operating buildings and to increase the international competitiveness of the U.S. building industry by providing modeling, measurement, and test methods needed to use advanced computation and automation effectively in construction, improve the quality of the indoor environment, and improve performance of building equipment. Computer-integrated construction is an expanding field. Facilities include a passive solar test house, solar calorimeters, six single-room environmental test houses, and solar collectors. A large environmental chamber, 14.9 x 12.8 x 9.5 meters high, is capable of testing two-story houses under simulated environmental conditions. Its earth floor can be excavated as needed for construction. A wide range of environments is possible, and the large chamber has been used to test buildings, special structures, and equipment in extreme climatic conditions. A guarded hot plate for measuring insulation performance and a calibrated hot box for studies of roof and wall sections are available. Test buildings are used and field testing is also carried out. Lighting facilities include a spectroradiometer and indoor and daylighting laboratories, as well as field instrumentation. The Plumbing Research Laboratory is a five-story facility using high-speed, preprogrammed data acquisition to study the performance of water supply and waste drainage systems.

Building Materials Division The division attempts to reduce building costs and increase building quality by providing technical bases for selecting the most cost-effective materials. They provide a technical base for selecting cost effective materials for buildings, and for standards, although the work may be years ahead of the actual standards. Research is related to the prediction of the service life of building materials. The organic materials laboratory studies paints, coatings, and roofing materials, with an emphasis on image analysis to predict the service life of building materials. The inorganic materials laboratory does advanced work in studying the chemical and physical changes that occur when cement reacts with water, using mathematical models and aiming at predictive modeling of the lifetime of concrete. Image analysis is used to study the growth of rust spots under coatings. Equipment includes an X-ray defractometer, scanning electron microscope, spectrophotometers, thermal analysis equipment, and very precise calorimeters.

The Construction Materials Reference Laboratories which are a part of this division conduct inspections at commercial test laboratories and private companies, acting as a contract advisory service. About 30 of the division's staff who provide this service are guest researchers from ASTM and AASHTO.

DISTINCTIVE ATTRIBUTES: The center is the only comprehensive building research laboratory in the United States. Several of the center's facilities are notable as either being the largest in the world, or the most precise or universal in capabilities of testing and measurement. The large-scale structural testing apparatus and the tri-directional test facility, used in earthquake testing; the large environmental chamber; and the hot box and hotplate, used to develop test standards, are examples.

PUBLICATIONS: Publishes its own reports, project summaries, and lists of publications, including the *Building Science Series*.

CENTER FOR FIRE RESEARCH

National Institute of Standards and Technology
Room A247-Polymers Building
Gaithersburg, MD 20899
301/975-6850

CONTACT: Jim Winger, Deputy Director

MISSION AND FOCUS OF RESEARCH: The center is engaged in the development of standard test methods to evaluate the physics and chemistry of fire, and determine the objective criteria for fire hazards using computer models. As a result of the technical work done at the center, local, state, and federal standards and codes for fire are established and reviewed regularly. The center also evaluates technologies for use in fire suppression and extinguishment, and also for lessening the effects and impacts of smoke and toxic gases.

DISTINCTIVE ATTRIBUTES: The center attempts to pull together all aspects of fire hazard assessment and studies of degradation of polymers in fire.

PUBLICATIONS: Professional and trade publications, NST reports, and a variety of government reports.

NAVAL CIVIL ENGINEERING LABORATORY

U.S. Navy
LO3C
Port Hueneme, CA 93043
805/982-4520

CONTACT: Robert Storer, Technical Director

MISSION AND FOCUS OF RESEARCH: NCEL conducts research on shore and offshore facilities for the Navy and Marine Corps with an annual budget of approximately $50 million. Building-related research includes work on physical security; nondestructive testing; development of alternative coatings to protect wood and metal from corrosive environments; paint, roofing, and pavement materials research; ventilation and cooling of buildings; passive solar structures; building thermal diagnostics; energy control systems; and design criteria for buildings under all types of loadings.

    The facilities include an advanced energy utilization test bed; an applied mechanics laboratory with simulated shock and vibration facilities, as well as a wind tunnel; an electric power laboratory; an optical metrology laboratory to conduct optical mechanics studies of stress, strain, and deflection; a soil mechanics laboratory for evaluation of soil mechanics, foundations and pavements; a materials laboratory to investigate organic coatings, plastics, metals, alloys, concrete, composite materials, and chemical problems related to environmental protection; and a pavement-loading facility that can apply loads up to 100,000 pounds to pavement in order to determine load-carrying capacities.

DISTINCTIVE ATTRIBUTES: NCEL contains one of the largest pressure chambers in the world for testing structures under high pressure. The laboratory has on staff some of the best experts in the nation on blast effects and has extensive capabilities for subjecting structures to blasts.

PUBLICATIONS: Technical journals, technical reports, and Tech Data sheets.

COLD REGIONS RESEARCH AND ENGINEERING LABORATORY

U.S. Army Corps of Engineers
72 Lyme Road
Hanover, NH 03755
603/646-4100

CONTACT: Andrew Assur, Chief Scientist

MISSION AND FOCUS OF RESEARCH: CRREL studies the characteristics of cold regions and applies this knowledge to the improvement of the living and work environments of people in cold climates. The main laboratory of CRREL houses 24 room-size refrigerated laboratories, many capable of achieving temperatures of $-30^{\circ}C$. An ice engineering research facility studies problems caused by ice in waterways. A newly completed frost effects research facility studies frost action in soils and below-freezing testing of pavements, foundations, and underground utilities.

DISTINCTIVE ATTRIBUTES: One of the few laboratories with extensive facilities focused on cold regions research.

CONSTRUCTION ENGINEERING RESEARCH LABORATORY (CERL)

U.S. Army Corps of Engineers
P.O. Box 4005
Champaign, IL 61820
800/872-2375

CONTACT: D. P. Mann, Information Management Office
Gilbert Williamson, Energy Systems Division
Robert Quattrone,
 Engineering and Materials Division
Ravinder Jain, Environmental Division
Edward Lotz, acility Systems Division

MISSION AND FOCUS OF RESEARCH: CERL conducts research to support the Army's military construction mission. Major research areas include new engineering practices and materials for construction, energy conservation and management, conservation of the environment, and the use of computers for managing the building resources at Army installations.
 CERL works through four divisions: (1) the Facilities Systems Division, including computer-aided design and specifications work; (2) Engineering and Materials Division,

including concrete mixtures and underground corrosion work; (3) Environmental Division, including noise standards; and (4) Engineering Systems Division, including computer simulations for buildings. CERL has a staff of 660 and a budget of $40 million of research; it is one of four major corps laboratories in the United States.

DIRECTORATE OF ENGINEERING AND SERVICES

U.S. Department of the Air Force
HQ USAF/L-Pentagon
Washington, DC 20330-5130
202/697-7366

CONTACT:	Joseph A. Ahearn, Director
	J. B. Cole, Associate Director

MISSION AND FOCUS OF RESEARCH: The directorate is responsible for the planning, design, and construction of Air Force facilities worldwide, including a wide variety of facilities such as operational, administrative, religious, educational, recreational, industrial, housing, commissaries, exchanges, utility systems, etc. The directorate develops and issues broad program goals and guidance to field offices, which are responsible for the daily program development and execution.

PUBLICATIONS: Air Force manuals, regulations, and pamphlets.

NAVAL FACILITIES ENGINEERING COMMAND

U.S. Department of Navy
Engineering and Design Criteria
Management Division
200 Stovall St.
Alexandria, VA 22332
703/325-0032

CONTACT:	Harry Zimmerman, Assistant Commander
	for Engineering and Design

MISSION AND FOCUS OF RESEARCH: The division develops and reviews architectural and engineering policies, criteria, and practices for the economical design and construction of shore facilities and fixed ocean structures to satisfy the functional/operational requirements in the best manner possible. The division also provides standard drawings and specifications, and directs and reviews all engineering and design efforts. Design and control of new and emerging technologies from project inception are other responsibilities of the division.

ENGINEERING AND SERVICES CENTER

U.S. Air Force
Tyndall Air Force Base, FL 32403
904/283-6310

CONTACT: James R. Van Orman, Deputy Director

MISSION AND FOCUS OF RESEARCH: Primary research efforts are directed toward special requirements for blast and shock resistance for U.S. Air Force buildings and facilities, including novel, protective structures for durability, blast, and penetration resistance; noise and sonic boom effects; and basic security, involving building components such as windows. The security of energy sources and the provision of redundant and self-contained sources, as well as of radioluminescent lighting, are of interest. Fire protection is an additional research concern. Environmental quality that relates to special Air Force requirements is also a subject of research. The total annual budget is $30 million with $3 million allocated to structures research.

DISTINCTIVE ATTRIBUTES: Remote site generation, use of energy, and exploration of uses of radioluminescence.

PUBLICATIONS: Defense Technical Information Center.

U.S. ARMY WATERWAYS EXPERIMENT STATION

Structures Laboratory
U.S. Army Corps of Engineers
P.O. Box 631
Vicksburg, MS 39180
601/634-3264

CONTACT: Bryant Mather, Chief,
Structures Lab, CEWES-SV-Z

MISSION AND FOCUS OF RESEARCH: Major activities of the Structures Laboratory include: research and development concerned with the behavior of concrete materials, elements, and structures; repair of concrete structures under in-use conditions; the resistance of such structures to dynamic forces such as earthquakes; effects of nuclear and conventional weapons detonated aboveground and underwater; design of protective structures and determination of their vulnerability; use of explosive technology for countermobility and mine field clearing; behavior of earth and rock subjected to intense transient loading; development of constitutive property definitions; and mathematical models to simulate behavior of geological materials.

DISTINCTIVE ATTRIBUTES: Dynamic force testing.

PUBLICATIONS: Defense Technical Information Center

NAVAL EXPLOSIVE ORDNANCE DISPOSAL TECHNOLOGY CENTER

U.S. Navy
Indian Head, MD 20640
301/743-4439

CONTACT: G. Burt Stephenson, Associate Technical Director

MISSION AND FOCUS OF RESEARCH: The center addresses matters relating to explosive devices, explosive effects, and countermeasures. The question of damage mitigation in relation to buildings is also addressed.

DISTINCTIVE ATTRIBUTES: The consideration of building design in relation to the potential effects of explosives.

FOREST PRODUCTS LABORATORY
(Forest Service)

U.S. Department of Agriculture
One Gifford Pinchot Drive
Madison, WI 53705
608/264-5600

CONTACT: Erwin L. Schaffer, Assistant Director, Wood Products Research

MISSION AND FOCUS OF RESEARCH: The Laboratory works to ensure the most efficient use of wood and wood resources. With regard to the use of wood products in buildings, research emphasizes using materials more effectively, improving structural integrity, increasing energy efficiency and developing more fire-safe products and structures. Research on the properties, design and performance of engineered wood structures and components is also performed. A current objective is the discovery and development of new concepts and procedures for preserving wood from biodegradation. Composite products and adhesives are studied. Future research will be shaped by the characterization of the chemical, physical, and mechanical properties of adhesives during and after bonding, and of bonded structures during exposure in service.
  The laboratory is operated and maintained in cooperation with the University of Wisconsin, where it is housed in 10 buildings on 22 acres; employs a staff of 300, 100 of whom are scientists and technical professionals; and has an annual budget of approximately $15 million.

LAWRENCE BERKELEY LABORATORY, CENTER FOR BUILDING SCIENCES

U.S. Department of Energy
University of California
Berkeley, CA 94720
415/486-4834

CONTACT: Arthur H. Rosenfeld, Director

MISSION AND FOCUS OF RESEARCH: The center's major objective is to investigate ways of reducing energy consumption in buildings. The center coordinates 300 staff members and $12.5 million in an effort that includes research on energy analysis, indoor environments, solar energy and windows, and daylighting. Emphasis is on research that will transfer quickly into the commercial market. LBL's energy analysis program continues to improve its DOE-2 computer model for predicting energy use in buildings. Other facilities include a mobile infiltration/test unit, a mobile window thermal test facility (MoWiTT), a room-size environmental chamber for studying emissions from consumer products, a sky simulator for studying different arrangements of windows to maximize building light without decreasing thermal efficiency, and a lighting technology lab for testing output and energy use of building lamps. A complete set of portable air quality monitoring equipment is also available.

DISTINCTIVE ATTRIBUTES: The center has the ability to instrument buildings for research into many different aspects of energy performance.

PUBLICATIONS: Technical journals, books, and in-house reports.

OAK RIDGE NATIONAL LABORATORY, BUILDINGS RESEARCH PROGRAM

Oak Ridge National Laboratory
Building 4500N, MS-188
Oak Ridge, TN 37831
615/574-5204

CONTACT: Roger Carlsmith, Director of Conservation and Renewable Energy Programs

MISSION AND FOCUS OF RESEARCH: ORNL works on improving energy efficiency in new and existing buildings in five major areas: building equipment, especially in developing more efficient heat pumps and appliances; roofs, especially flat or low-slope roofs for commercial buildings, as well as analysis and development of energy, mechanics, and maintenance aspects; thermal envelopes, including the study of thermal anomalies, thermal mass considerations, and diagnostic procedures of energy efficiency; retrofit; and conservation, including responsibility for the Residential Conservation Service (RCS), which provides information on technical issues to utilities.

Facilities include a roof test facility and an indoor-outdoor environmental chamber. Three test buildings at the University of Tennessee, Knoxville, the Tennessee Energy Conservation in Housing (TECH), are used for the investigation of building equipment. Retrofit options are tested in three houses in the Karns community, a subdivision near Oak Ridge.

The end-use efficiency studies make up about 10 percent of the Department of Energy's budget for ORNL. Much of the research work is subcontracted to industry on a cost-sharing percentage basis.

DISTINCTIVE ASPECTS: Major improvements in the technology of gas-fired heat pumps have made them more promising for greatly increased residential and commercial use.

PUBLICATIONS: Topical reports on which abstracts are available.

BATTELLE-PACIFIC NORTHWEST LABORATORIES

U.S. Department of Energy
P.O. Box 999
Richland, WA 99352
(509) 375-4359

CONTACT: R. William Reilly, Director, Energy Systems Department

MISSION AND FOCUS OF RESEARCH: Battelle's emphasis is on the evaluation of energy use in commercial and industrial buildings. Energy performance monitors that use low-cost data-logging instruments are used to determine energy use and air quality. These instruments and computer analyses help determine the effects of building changes and air exchange rate reductions. Work is done on new building design for energy efficiency as well as on retrofit. Mobile homes are also monitored. The Department of Energy maintains a $100-million laboratory facility and Battelle Memorial Institute supports $45 million in laboratory work.

PUBLICATIONS: Technical journals.

SOLAR ENERGY RESEARCH INSTITUTE

U.S. Department of Energy
1617 Cole Boulevard
Golden, CO 80401
303/231-7115

CONTACT: Steve Rubin, Technical Inquiries Service
Larry Flowers, Technical Program Integrator
Tom Potter, Leader, Conservation Programs

MISSION AND FOCUS OR RESEARCH: Serving as the national center for solar energy research, the institute maintains more than 50 specialized laboratories and test facilities for research by the private sector and universities into various components of production, storage, and uses of solar energy in a variety of applications. Laboratories aimed at building research include the thermal analysis laboratory, the cooling laboratory, the low temperature heat/mass transfer laboratory, and a materials laboratory that is involved with elements such as glass and various approaches to glazing.

PUBLICATIONS: All publications are available through U.S. Government Printing Office and National Technical Information Center.

WATER ENGINEERING RESEARCH LABORATORY

U.S. Environmental Protection Agency
Cincinnati, OH 45268
513/569-7509

CONTACT: Roger Wilmuth, Chief, Manufacturing and Services Industries Bureau

MISSION AND FOCUS OR RESEARCH: The lab's emphasis is on the efficient and effective removal of asbestos with minimum environmental impact, which includes the evaluation of control methods. Currently there is a staff of four, with a budget of $500,000 in agency funds.

DISTINCTIVE ATTRIBUTES: The only group evaluating control technologies for asbestos.

PUBLICATIONS: Available through National Technical Information Service.

TAFT LABORATORY

National Institute for Occupational
Safety and Health (NIOSH)
U.S. Department of Health and Human Services
4676 Columbia Parkway
Cincinnati, OH 45226
513/841-4221

CONTACT: James Gideon, Chief, Engineering Control Technology Branch

MISSION AND FOCUS OF RESEARCH: Taft Laboratory evaluates control techniques and technologies used for the removal of asbestos. Agency funds supply a budget of $100,000, with an additional $25,000 from the Environmental Protection Agency.

DISTINCTIVE ATTRIBUTES: Use of electron microscopy and aggressive sampling to determine clearance.

PUBLICATIONS: NIOSH and NTIS technical reports are used, as well as the *American Industrial Hygiene Association Journal* and the *Applied Industrial Hygiene Journal*.

OTHER PROFILES

BATTELLE MEMORIAL INSTITUTE

505 King Avenue
Columbus, OH 43201
614/424-6476

CONTACT: John R. Hagely, Project Manager and Senior Research Architect

MISSION AND FOCUS OF RESEARCH: Battelle is a science-based, internationally recognized research and development organization, generating contracts with private industry; local, state, and federal governments; and other nations.

Building-related research includes new building material technologies; coatings testing and analysis; upgrading thermal performance of glass; evaluation of institutional barriers for new building systems; design trend analysis in industry R&D laboratories; life-cycle cost analysis of building systems and building materials, including economic and technical aspects; the development of global building profiles to analyze traditional building materials used around the world; design and analysis of tools and equipment used in buildings; testing of physical and mechanical properties of insulation; and test problems of adhesive-bonded roof trusses. Research in the energy division includes improving the efficiency of burner systems and furnaces; and new techniques and materials for photovoltaic cells.

The institute has a staff of 3,500 people, half of whom are technical staff. It averages 2,400 projects a year and has an annual budget of about $100 million per year.

DISTINCTIVE ATTRIBUTES: Unique in size and scope of effort. Facilities are all in-house. The institute has worldwide contacts and facilities.

PUBLICATIONS: Annual reports, papers, articles, and books.

IIT RESEARCH INSTITUTE

10 West 35th Street
Chicago, IL 60616
312/567-4000

CONTACT: Andrew D. Farrell, Director of Government Programs, Washington, D.C.
202/296-1610

MISSION AND FOCUS OF RESEARCH: Major areas of building-related research include fire, explosive mechanics, materials, and indoor air quality. The institute has a fire test laboratory to test fire spread and the effects of fire, which is also capable of testing the contents of buildings to full-scale buildings. Analysis and computer modeling are used in fire research. Projects in the explosive mechanics laboratory include the manufacturing and testing of explosives in confinement. There is also testing of the effects of explosives on structures and on materials. Materials laboratory work focuses on concrete, granite, stone, erosion, the strengthening of building materials, and coatings. Indoor air quality projects measure what is in the air and analyze causes and effects of indoor air problems. The chemistry of the indoor environment, the types of organic components present in the air, desiccants, and the effects of unvented gas appliances on indoor air are studied.

The budget allocated for research in these areas is between $2 and $5 million. Sources include the Environmental Protection Agency, the National Institutes of Health, the Department of Defense, the Federal Emergency Management Administration, and private industry.

PUBLICATIONS: Reports of project activities written to sponsors and journals.

NATIONAL CENTER FOR APPROPRIATE TECHNOLOGY

3040 Continental Drive
P.O. Box 3838
Butte, MT 59702
406/494-4572

CONTACT: Diane Smith, Public Information Manager

MISSION AND FOCUS OF RESEARCH: NCAT's objectives are to document effective energy conservation techniques and building technologies, and to transfer that information to consumers, government agencies, and building professionals. Recent work includes end-use load monitoring of commercial and residential buildings and documenting the performance of superinsulated homes. Funding sources include state and federal agencies, national labs, and utilities.

DISTINCTIVE ATTRIBUTES: A private, nonprofit agency created by Congress to transfer technology, the center stresses the wise use of resources and environmental safety.

PUBLICATIONS: Own publications.

OTHER SERVICES: Educational, informational.

NATIONAL INSTITUTE OF BUILDING SCIENCES

1201 L St., N.W.
Suite 400
Washington, DC 20005
202/289-7800

CONTACT: Don Hill, Director of Communications

MISSION AND FOCUS OF RESEARCH: The institute is a 10-year-old private, nonprofit organization authorized by Congress with the mission of enhancing the building regulatory environment and encouraging the introduction of new and existing technologies into the building process.

RESEARCH TRIANGLE INSTITUTE

P.O. Box 12194
Research Triangle Park, NC 27709
803/334-8571

CONTACT: Reid Maness, Management Support Office

MISSION AND FOCUS OF RESEARCH: Building-related research focuses on environmental testing of toxic substances used in building materials; on other environmental problems such as asbestos, indoor air pollution, and radon contamination; and on energy conservation economics and engineering. The institute is housed in Research Triangle Park in a complex of 15 laboratory and office buildings. Facilities include laboratories for chemistry, environmental chemistry, toxicology, histology, chemical engineering, and polymers, as well as computer support for economics and statistical studies. The federal government provides about 90 percent of the institute's support.

SOUTHWEST RESEARCH INSTITUTE

6220 Culebra Road
San Antonio, TX 78284
512/684-5111

CONTACT: Ulric S. Lindholm, Vice President,
Division of Engineering and Materials Sciences
Gordon E. Hartzell, Director,
Department of Fire Technology

MISSION AND FOCUS OF RESEARCH: The institute's Department of Fire Technology conducts a wide variety of standard fire tests for flame spread, fire endurance, flammability, and smoke toxicity. Custom-designed tests are developed to determine the fire performance characteristics of innovative systems or to replicate specific fire scenarios of interest. Third-party certification, listing and labeling, with follow-up inspection of products, components and materials, are provided by the institute.

The institute has 1,000 or more research programs, which primarily serve the power generation, offshore, transportation and aerospace industries. Research efforts of interest to the building community include work on blast resistance of structures; on the effectiveness of controlled-access barriers; geotechnical engineering measurement of ground stability using elastic wave propagation between two boreholes; and seismic studies, including the use of two shake tables.

The institute is an independent, nonprofit research institution on 765 acres with more than 2,000 employees, of whom about 800 are professional, 800 technical support, and the remainder administrative support. About 60 percent of its research earnings, which were $140 million in 1986, comes from private industry, with the remainder from government contracts.

DISTINCTIVE ATTRIBUTES: In fire technology, the institute has three high-bay test areas totaling 9,400 square feet for large-scale testing. Complete furnace facilities are available for determining fire endurance of assemblies in both horizontal and vertical configurations.

PUBLICATIONS: *The Journal of Fire Sciences.*

SRI INTERNATIONAL

333 Ravenswood Avenue
Menlo Park, CA 94025
415/859-5820

CONTACT: Kay Clark, Director of Corporate Relations

MISSION AND FOCUS OF RESEARCH: SRI International is a private and independent professional organization that serves the public interest by providing a broad range of basic and applied research, advisory, and technical services worldwide. Research related to the building industry includes the improvement and development of new materials, polymer sciences, energy efficiency, corrosion research, and analysis of structural failures. There is a research staff of 3,000 people and a budget of $210 million with 2 to 3 percent allocated to the building industry. Sources for the budget are industry, federal, state, and local governments.

PUBLICATIONS: An annual report, journal and newspaper articles, documentations from centers, and news releases.

II

INTERNATIONAL

ARGENTINA

ARGENTINA PORTLAND CEMENT INSTITUTE

San Martin 1137
1004 Buenos Aires

CONTACT: Carlos Ernesto Duvoy, Technical Director

MISSION: To represent the cement industry in promoting, developing, and applying portland cement technology.

PRIMARY WORK: Performs research through six divisions: (1) roads, (2) structures, (3) home building, (4) prefabrication, (5) prestressed concrete, and (6) technological development. Also has information dissemination activities and conducts courses and seminars.

SOURCE OF FINANCES: Cement industry.

PUBLICATIONS: Technical publications, bulletins, annual bibliographies, and library catalogues.

AUSTRALIA

AUSTRALIAN INSTITUTE OF STEEL CONSTRUCTION, LTD.

110 Alfred Street
P.O. Box 434, Milsons Point 2061
New South Wales

CONTACT: I. R. Hooper, Chief Executive

MISSION: To serve the steel construction industry by promoting the use of steel in all forms of construction.

PRIMARY WORK: Provides technical advisory services to members; organizes technical conferences, seminars, and conventions; conducts student courses; provides library and bookshop services; informs government, educators, and decision makers with respect to the use of steel in construction.

SOURCE OF FINANCES: Membership fees and sales of publications.

PUBLICATIONS: Quarterly journals and specialty publications.

AUSTRALIAN ROAD RESEARCH BOARD (ARRB)

P.O. Box 156, Bag 4
Nunawading 3131, Victoria

CONTACT: M. G. Lay, Director

MISSION: National laboratory for road research and information.

PRIMARY WORK: Performs research on issues of road and traffic engineering, transport planning, and instrumentation; maintains national documentation center.

SOURCE OF FINANCES: Commonwealth and state governments; consultation fees.

PUBLICATIONS: Quarterly journal and index, research and special reports, technical manuals, and annual progress report.

BRICK DEVELOPMENT RESEARCH INSTITUTE (BDRI)

159 Drummond Street
Carlton South
Victoria 3053

CONTACT: Tony Larnach-Jones, Executive Director
Clive Letts, Research Manager

MISSION: To represent the Australian and New Zealand clay brick and paver industry in technical matters; to carry out research and testing; and to provide advice to manufacturers, specifiers, and users of clay bricks and pavers.

PRIMARY WORK: Advisory services through inspection, research, testing, publications and design.

SOURCE OF FINANCES: Member subscriptions, fees from consulting, publications sales, and grants from the federal government.

PUBLICATIONS: Technical notes, design guides, specifications, and test methods for clay bricks and pavers in written and video form.

CEMENT AND CONCRETE ASSOCIATION OF AUSTRALIA

147 Walker Street
North Sydney, New South Wales, 2060

CONTACT: K. J. Cavanagh, Director

MISSION: To promote and develop the use of cement and concrete.

PRIMARY WORK: Provides technical advisory services, funds research on cement and concrete, develops bases for national standards, and promotes new uses of cement and concrete.

SOURCE OF FINANCES: Portland cement industry.

PUBLICATIONS: Journals and periodic technical manuals, reports, notes, and information sheets.

COMMONWEALTH SCIENTIFIC AND INDUSTRIAL RESEARCH ORGANIZATION, DIVISION OF BUILDING RESEARCH (CSIRO, DBR)

P.O. Box 56, Highett
Victoria 3190

CONTACT: F. A. Blakey, Chief

MISSION: National research laboratory seeking to increase efficiency and effectiveness in the building and construction sector of the economy; to enhance the potential standard of the infrastructure for all Australians at work, at play, and at home; to minimize any adverse impacts of the construction sector on the environment; and to collaborate with and give advice to industry and the community.

PRIMARY WORK: Performs research on building technology, civil engineering works, acoustics, housing, and social and economic studies related to the infrastructure.

SOURCE OF FINANCES: Government and private industry.

PUBLICATIONS: Bimonthly newsletter, annual research papers, and research reports.

CONCRETE INSTITUTE OF AUSTRALIA

25 Berry Street
North Sydney, New South Wales 2060

CONTACT: K. J. Cavanagh, Secretary/Treasurer

MISSION: To develop and disseminate concrete technology and practices.

PRIMARY WORK: Hosts conventions and meetings; publishes technical papers.

SOURCE OF FINANCES: Membership fees.

PUBLICATIONS: Quarterly journal, reference specifications for concrete work, technical committee reports, and technical papers.

QUEENSLAND INSTITUTE OF TECHNOLOGY, FACULTY OF THE BUILT ENVIRONMENT, DEPARTMENT OF ARCHITECTURE AND INDUSTRIAL DESIGN

G.P.O. Box 2434
Brisbane
Queensland

CONTACT: Bill B. P. Lim, Head

MISSION: To provide academic education and research to the building environment and architectural design and practice.

PRIMARY WORK: Research is performed in architectural design and planning, lighting, energy conservation and renewal, and thermal behavior.

SOURCE OF FINANCES: Government.

UNIVERSITY OF NEW SOUTH WALES, SCHOOL OF BUILDING

P.O. Box 1, Kensington
New South Wales 2033

CONTACT: Roger M. A. Miller, Head

MISSION: To provide academic education and technical assistance for Australia's building industry.

PRIMARY WORK: Research methods of building construction, project management, contract preparation, and administration; evaluates building failures; has close working relationships with the Australian building community; and provides knowledge about technical and managerial problems in building construction.

SOURCE OF FINANCES: University of New South Wales, Australian government, and building and materials manufacturers.

PUBLICATIONS: Research results and specialty publications.

NEW SOUTH WALES UNIVERSITY, SCHOOL OF CIVIL ENGINEERING

Kensington
New South Wales 2033

CONTACT: D. G. Carmichael, Director

MISSION: To provide academic education, perform research in civil engineering construction and management, and furnish consultative services for the public and private building industry.

PRIMARY WORK: Performs evaluation of field construction operations, concrete placement systems, and equipment management; and performs studies on building construction techniques and material placement.

SOURCE OF FINANCES: University and sponsored research.

PUBLICATIONS: Research papers.

UNIVERSITY OF SYDNEY, DEPARTMENT OF ARCHITECTURAL SCIENCE

New South Wales 2006

CONTACT: John S. Gero, Director

MISSION: To perform teaching and research for improving building design and performance.

PRIMARY WORK: Performs work on computer-aided design, knowledge-based systems, acoustics, lighting, thermal environment, building services, building materials, and computer applications to building and architecture.

SOURCE OF FINANCES: Government and industry.

PUBLICATIONS: Journal papers, books, research reports, quarterly reviews, and biannual newsletters.

AUSTRIA

RESEARCH SOCIETY FOR HOUSING, BUILDING, AND PLANNING

Lowengasse 47
P.O. Box 164, A 1031 Wien

CONTACT: Ewald Liepolt, Director

MISSION: To perform interdisciplinary housing research.

PRIMARY WORK: Conducts research in housing, building, and planning.

SOURCE OF FINANCES: Membership fees, sales of publications, and sponsored research.

PUBLICATIONS: Monographs, serial publications, working papers, and lists of publications.

AUSTRIAN INSTITUTE FOR BUILDING RESEARCH

An den langen Lussen A
1190 Wien

CONTACT: Michael Wachberger, Scientific Director

MISSION: Central building technology research institute.

PRIMARY WORK: Performs work in rehabilitation, energy conservation, and organization and management of construction.

SOURCE OF FINANCES: Member fees, government, and fees for services.

PUBLICATIONS: Research reports, annual report, and news bulletin.

AUSTRIAN CEMENT RESEARCH INSTITUTE

Reisnerstrasse 53
A 1030 Wien

CONTACT: Hermann Sommer, Director

MISSION: To serve as the research institute for Austria's cement industry.

PRIMARY WORK: Performs research on cement and concrete, developments of new applications, quality control of cement and other hydraulic binding agents, and disseminates information.

SOURCE OF FINANCES: Cement industry.

PUBLICATIONS: Quarterly journal, bulletins, and annual lists of publications.

LABORATORY FOR PLASTICS TECHNOLOGY

Wexstrasse 19-23
1200 Wien

CONTACT: H. Hubney, Director

MISSION: To serve as a plastic testing and research laboratory.

PRIMARY WORK: Performs research on plastics and building technology; develops training courses in plastics technology.

SOURCE OF FINANCES: Membership fees and sponsored projects.

PUBLICATIONS: Reports published in scientific journals.

AUSTRIAN WOOD RESEARCH INSTITUTE

Arsenal, Franz-Grill-Strasse 7
A 1030 Wien

CONTACT: Herbert Neusser, Scientific Director

MISSION: To serve the Austrian Wood Research Society for chemical and mechanical technologies of wood.

PRIMARY WORK: Performs work in the mechanical technology of wood and composite woods and the chemical technology of wood, such as glues, varnishes, pulp, and paper.

SOURCE OF FINANCES: Sponsored research, services for industry, and member fees.

PUBLICATIONS: Bimonthly journal and quarterly card index service.

AUSTRIAN PLASTICS INSTITUTE

Arsenal, Objekt 213
Franz-Grill-Strasse 5
A 1030 Wien

CONTACT: Otto Hinterhofer, Director

MISSION: To serve as a private research and testing institute.

PRIMARY WORK: Performs basic and applied research and testing of plastics and composites, including fiber-reinforced plastics and plastic concrete.

SOURCE OF FINANCES: Fees for research and industrial services.

PUBLICATIONS: Specialty publications.

PHYSICAL-TECHNICAL INSTITUTE FOR RESEARCH ON HEAT AND SOUND TECHNOLOGY

Wexstrasse 19-23
A 1200 Wien

CONTACT: Judith Lang, Director

MISSION: To serve as a research and testing laboratory for the building community.

PRIMARY WORK: Heat and sound insulation in buildings, building physics, noise control (machinery noise, traffic noise), and room acoustics.

SOURCE OF FINANCES: Government.

PUBLICATIONS: Research results published in various journals and serials.

TECHNICAL UNIVERSITY OF GRAZ--INSTITUTE FOR SOIL MECHANICS, ROCK MECHANICS, AND FOUNDATION ENGINEERING

Rechbauerstrasse 12
A 8010 Graz

CONTACT: M. Fuchsberger, Director

MISSION: Education and research in engineering mechanics.

PRIMARY WORK: Performs research in foundations, soil and rock mechanics, soil testing, and large-scale field testing.

SOURCE OF FINANCES: Government and fees for services.

PUBLICATIONS: Periodic journals and research reports.

TECHNICAL UNIVERSITY OF VIENNA, INSTITUTE FOR BUILDING CONSTRUCTION

Karlsplatz 13
A 1040 Wien

CONTACT: Alfred Pauser

MISSION: Education and research to support the building community.

PRIMARY WORK: Research is performed on civil engineering construction, building physics and acoustics, and building information systems.

SOURCE OF FINANCES: Government.

PUBLICATIONS: Biannual report and quarterly director's report.

INSTITUTE FOR BUILDING MATERIALS, TECHNICAL INDUSTRIAL MUSEUM

Wexstrasse 19-23
A 1200 Wien

CONTACT: Gerhard Kraml, Chief

MISSION: Testing of construction materials and equipment.

PRIMARY WORK: Performs tests on building materials and construction systems.

SOURCE OF FINANCES: Government.

PUBLICATIONS: Reports of tests and annual report.

BELGIUM

TECHNICAL CONTROL BUREAU FOR CONSTRUCTION SAFETY (SECO)

Rue d'Arlon 53
1040 Brussels

CONTACT: D. Vandepitte, Chairman of the Technical Board
J. R. Richelle, Managing Director

MISSION: To provide quality control to improve the design and construction of buildings and the manufacture and use of building materials in order to reduce the risk of defects.

PRIMARY WORK: Performs quality control for building construction and civil engineering works, building services, and manufactured materials; and performs research to improve quality control of constructions.

SOURCE OF FINANCES: Fees from projects.

PUBLICATIONS: Research report, technical reviews, and conference and seminar reports.

BELGIUM CENTER FOR CORROSION STUDY (CEBELCOR)

Avenue Paul Heger, grill 2
B 1050 Brussels

CONTACT: Antoine Pourbaix, Director

MISSION: To perform corrosion research; to provide consulting, testing, and educational services.

PRIMARY WORK: Research is performed in metallic corrosion, electrochemistry, mechanisms of corrosion, behavior of materials, steel in concrete, and water supply systems.

SOURCE OF FINANCES: Research contracts and consulting and member fees.

PUBLICATIONS: Periodic research reports and conference proceedings.

NATIONAL CEMENT RESEARCH CENTRE (CRIC)

Rue Cesar Franck 46
1050 Brussels

CONTACT: J. Urlings, President

MISSION: To serve as an industrial technical center for the cement industry.

PRIMARY WORK: Performs research on concrete and cement with a testing and technical laboratory.

SOURCE OF FINANCES: Research grants, contracted work, and cement manufacturers fees.

PUBLICATIONS: Research reports, activity reports, and monthly issue of CRIC document card systems.

BELGIAN BUILDING RESEARCH CENTER (CSTC)

Rue du Lombard 41
B 1000 Brussels

CONTACT: Ir Carlo DePauw, Director General

MISSION: National industrial research center for building research, documentation, and technical assistance.

PRIMARY WORK: Performs research on building construction, civil engineering, building performance, energy conservation, and environmental issues; provides technical assistance to contractors and manufacturers; and conducts training and seminars.

SOURCE OF FINANCES: Levy on contractors, government, and contracts.

PUBLICATIONS: Quarterly journal, technical notes, research reports, weekly paper, and other specialty publications.

COATINGS RESEARCH INSTITUTE (CoRI)

Avenue P. Holoffe
B 1342 Limelette

CONTACT: P. Janssen Bennynck, Managing Director

MISSION: To promote scientific development and technical progress in the areas of coatings, paints, and inks.

PRIMARY WORK: Performs research, analysis, testing, and assessments of anticorrosion, industrial paints, and new technologies that result in specifications; provides technical assistance, documentary services, and bibliographic studies; offers scientific and technical training and seminars.

SOURCE OF FINANCES: Public and private.

PUBLICATIONS: Research reports, information bulletins, and international scientific journals.

NATIONAL GLASS INSTITUTE (INV)

Boulevard Defontaine 10
B 600 Chaleroi

CONTACT: M. Verburgh, Director

MISSION: To promote the development of basic and applied research for the Belgian glass industry.

PRIMARY WORK: Research and testing for the Belgium glass manufacturers in thermo-physics of glass making, physical and chemical properties of glass, fusion and refining production of glass, and special problems associated with mirrors, automotive glass, and double glazing.

SOURCE OF FINANCES: Belgian glass industry, and sales of services.

PUBLICATIONS: Research reports and abstracts.

CATHOLIC UNIVERSITY OF LEUVEN, DEPARTMENT OF ARCHITECTURE

Karsteel Arenberg
Kardinal Mercierlaan 94
B 3030 Heverlee

CONTACT: Jan Delrue, Department President

MISSION: Education, research, and testing.

PRIMARY WORK: Performs research in building physics, design and building process, preservation of historic districts and buildings, and technical support in areas such as the theory and history of architecture and urban and regional planning.

SOURCE OF FINANCES: University and government.

PUBLICATIONS: Specialized reviews and scientific reports.

CATHOLIC UNIVERSITY OF LEUVEN, CIVIL ENGINEERING DEPARTMENT

Park van Arenberg
de Croylaan 2
B 3030 Leuven
Heverlee

CONTACT: J. Berlamont, President

MISSION: To provide advanced education and fundamental and applied research on materials, structures, and construction.

PRIMARY WORK: Performs research in civil engineering; reinforced and prestressed concrete; structural restoration and strengthening masonry; concrete structures; improving the quality of roads; and modeling flows of rivers, channels, canals, and the North Sea.

SOURCE OF FINANCES: University, government, and fees for services.

PUBLICATIONS: Journal articles, student theses, and periodic reports.

CATHOLIC UNIVERSITY OF LEUVEN, ACOUSTICS AND HEAT CONDUCTION LABORATORY

Celestijnenlaan 200 D
B 3030 Heverlee

CONTACT: H. Myncke, Director

MISSION: Education and research on the acoustical environment.

PRIMARY WORK: Performs research in acoustics and noise abatement, heat conductivity, and water vapor transmission.

SOURCE OF FINANCES: Ministry of Education and sponsored research.

PUBLICATIONS: Scientific journals; periodic reports.

STATE UNIVERSITY OF GHENT, FACULTY OF ENGINEERING, AND LABORATORY FOR FUEL TECHNOLOGY AND HEAT TRANSFER.

St. Pieternieuwstraat 41
9000 Gent

CONTACT: R. Minne, Director

MISSION: To serve as university research laboratory for the building community.

PRIMARY WORK: Provides measurement techniques for heat transfer through building materials, fuel combustion technology, and fire research.

SOURCE OF FINANCES: University, contract research, and testing fees.

PUBLICATIONS: Research results and professional journals.

GHENT STATE UNIVERSITY, MAGNEL LABORATORY FOR REINFORCED CONCRETE

Grotesteenweg-Noord 2
B 9710 Ghent

CONTACT: H. Lambotte, Director

MISSION: Education, research, and testing of building materials.

PRIMARY WORK: Performs research on reinforced and prestressed concrete and other building materials; undertakes measurement and acceptance testing for industry.

SOURCE OF FINANCES: Government, research contracts, and fees for testing.

PUBLICATIONS: Journals.

CATHOLIC UNIVERSITY OF LOUVAIN, CENTER FOR ARCHITECTURAL RESEARCH (CRA)

Batiment Vinci
Place du Levant 1
B 1348 Louvain-la-Neuve

CONTACT: J. F. Mabardi, Director

MISSION: To educate and advance architectural practice through improved technology.

PRIMARY WORK: Performs research in architecture, town planning, building engineering, energy conservation, and alternate technologies.

SOURCE OF FINANCES: University, research contracts, and government.

PUBLICATIONS: Periodic research reports and professional journals.

STATE UNIVERSITY OF LIEGE, RESEARCH CENTERS FOR ARCHITECTURE AND TOWN PLANNING, CIVIL ENGINEERING, BUILDING PHYSICS LABORATORY

Quai Banning 6
4000 Liege

CONTACT: J. Englebert, F. Peters, G. Fonder, R. Maquoi, and A. Dupagne, Directors

MISSION: To educate and serve as a research center for architectural, structural, and building physics research.

PRIMARY WORK: Performs research in architecture and town planning, renovation and preservation of buildings and districts, building materials; materials use, structural engineering, computer-aided design, HVAC equipment, and thermal systems.

SOURCE OF FINANCES: University, industry, government, and research contracts.

PUBLICATIONS: Technical notes, bi-annual list of publications, lecture notes, periodic reports, and international journals.

STATE UNIVERSITY OF LIEGE, CIVIL ENGINEERING TEST LABORATORIES AND DEPARTMENT OF STRENGTH OF MATERIALS

Quai Banning 6
4000 Liege

CONTACT: A. Lejeune, President

MISSION: To provide education and research laboratory capabilities to the Belgian building industry.

PRIMARY WORK: Performs theoretical, numerical, and experimental research in civil engineering, including soil and rock mechanics, bridges, structural engineering, reinforced concrete, building materials, and hydraulic engineering.

SOURCE OF FINANCES: Research contracts, government, and grants.

PUBLICATIONS: University publications, professional journals, and periodic reports.

BRUSSELS FREE UNIVERSITY

Avenue Adolphe Buyl 87
B 1050 Brussels

CONTACT: P. Halleux and Jean Nuyens, Directors

MISSION: To provide education in civil engineering and materials testing.

PRIMARY WORK: Performs research on structures, soils, rock, and fracture mechanics; materials testing.

SOURCE OF FINANCES: University, research contracts, and grants.

PUBLICATIONS: Research reports, student theses, and technical notes.

BRAZIL

NATIONAL COMMISSION FOR THE CIVIL CONSTRUCTION INDUSTRY (CNICC)

Rua Mariz e Barros
13-3 andar
Rio de Janeiro, 20.270 RJ

CONTACT: Marcos Jose Marques, Secretary General of Ministry of Industry and Commerce

MISSION: To promote improved standards through advising the government about job programs, civil construction, and modernization of the construction industry.

PRIMARY WORK: Performs work in civil engineering and civil works, economics, documentation; delivery methods for social development concerning population; national information system on the construction market.

SOURCE OF FINANCES: Ministry of Industry and Commerce.

PUBLICATIONS: Quarterly information report and biannual statistical bulletin.

L. A. FALCAO BAUER TESTING LABORATORY

Rua Aquinos 111
05036 Sao Paulo, SP

CONTACT: L. A. Falcao Bauer, President

MISSION: Private laboratory testing services for the Brazilian building community.

PRIMARY WORK: Provides measurement, analyses, evaluation, and testing services in civil construction, mechanical, and chemical industries; quality assurance; calibration; failure analysis; and expert witness.

SOURCE OF FINANCES: Contracted work for testing and services.

PUBLICATIONS: Technical standards.

TECHNOLOGICAL CENTRE OF MINAS GERAIS (CETEC)

Av Jose Candido de Silveira 2000
Horto 300.000
Belo Horizonte
Minas Gerais

CONTACT: Olavo Machado Jr., President

MISSION: To improve industrial processes through technical assistance, applied research, analysis, and testing.

PRIMARY WORK: Provides services in architectural design; building technology and materials; low-cost building; environmental engineering; research and testing services; and studies the trends, economies, and availability of resources.

SOURCE OF FINANCES: Government and industrial contracts.

PUBLICATIONS: Bimonthly technical bulletins, notes, manuals, and proceedings.

TECHNOLOGICAL RESEARCH INSTITUTE OF THE STATE OF SAO PAULO (IPT)

Cidade Universitaria Armano Salles de Oliveria
Caixa Postal 7141
055008 Sao Paulo

CONTACT: Henrique Silveira de Almeida, Director Superintendent

MISSION: To advance building practices through the technical services of a government research laboratory.

PRIMARY WORK: Performs research in civil engineering, construction, building technology, materials, building physics, and structural loads; provides testing and analysis support; designs scientific equipment.

SOURCE OF FINANCES: State of Sao Paulo, government, contract services, and grants.

PUBLICATIONS: Technical notes, results of research, and annual lists of publications.

CANADA

CANADA MORTGAGE AND HOUSING CORPORATION

682 Montreal Road
Ottawa, Ontario K1A 0P7

CONTACT: D. A. Stewart, Director, Research Division

MISSION: To promote the construction of new houses, the repair and modernization of existing housing, and the improvement of housing and living conditions.

PRIMARY WORK: Plans and undertakes research directed at understanding the conditions and relationships affecting the supply and demand for housing in Canada, along with its quality and affordability; provides external incentives which support independent research and scholarship in the field of housing.

SOURCE OF FINANCES: Government.

PUBLICATIONS: Research reports, books, statistical publications, and bibliographies on housing and community development.

CANADIAN CONSTRUCTION MANAGEMENT INSTITUTE (CMI),
CANADIAN CONSTRUCTION MANAGEMENT DEVELOPMENT INSTITUTE (CCMDI)

5799 Yonge St.
Suite 901
Toronto, Ontario M2M 3V3

CONTACT: David Judge, President and Executive Director

MISSION: To coordinate and promote construction industry management development and training programs.

PRIMARY WORK: Provides link to industry and academia by coordinating management programs, consultation services, and information dissemination activities.

SOURCE OF FINANCES: Construction and housing associations and government.

PUBLICATIONS: Course/program catalogue, newsletter, and software directory.

CANADIAN INSTITUTE OF STEEL CONSTRUCTION (CISC)

201 Consumers Road
Suite 300
Willowdale, Ontario M2J 4G8

CONTACT: Hugh A. Krentz, President

MISSION: To provide technical and marketing services for improving the efficiency and economy of structural steel.

PRIMARY WORK: Develops and disseminates technology on steel construction, including computer design aids, computer programs, training, and information.

SOURCE OF FINANCES: Member fees.

PUBLICATIONS: Technical standards and recommendations, research reports, bibliographies, proceedings, recommended practices, and other specialty publications.

CENTER FOR BUILDING STUDIES

1455 de Maisonneuve Boulevard W
Montreal, Quebec H3G 1M8

CONTACT: Paul Fazio, Director

MISSION: To provide research and teaching services.

PRIMARY WORK: Performs research in building environment, building structures, construction management, and building structures; offers undergraduate and graduate studies in building engineering.

SOURCE OF FINANCES: University, government, and research contracts.

PUBLICATIONS: Proceedings and specialty publications.

CENTER FOR RESEARCH AND DEVELOPMENT IN MASONRY (CRDM)

105 4528 6A Street NE
Calgary, Alberta T2E 4B2

CONTACT: Emlyn L. Jessop, Director General

MISSION: To promote the advancement of masonry science and practice.

PRIMARY WORK: Performs masonry research, development, and testing for processes and field application.

SOURCE OF FINANCES: Consulting and testing fees, membership fees, and research grants.

PUBLICATIONS: Technical standards, research reports, bibliographies, and specialty publications.

CLAY BRICK ASSOCIATION OF CANADA (CBAC)

1 Sparks Avenue
Willowdale, Ontario M2H 2W1

CONTACT: John F. Cutler, Managing Director

MISSION: To promote the advancement of clay brick use.

PRIMARY WORK: Provides technical assistance and engineering support to manufacturers, represents member companies, and promotes quality assurance of clay brick.

SOURCE OF FINANCES: Member fees.

PUBLICATIONS: Specialty publications.

INSTITUTE FOR RESEARCH IN CONSTRUCTION, NATIONAL RESEARCH COUNCIL OF CANADA

Montreal Road
Ottawa, Ontario K1A OR6

CONTACT: George Seaden, Director

MISSION: To serve as the national building research laboratory to provide improved technology to the construction industry.

PRIMARY WORK: Performs research in fire performance; building materials, services, and structures; geotechnical engineering; noise and vibration; codes and standards; information systems; provides testing services through contract research; performs advisory services; undertakes technology transfer activities.

SOURCE OF FINANCES: Government and contract research.

PUBLICATIONS: Research reports, publication series, and specialty technical publications.

PUBLIC WORKS CANADA (PWC)

Ottawa, Ontario K1A 0M8

CONTACT: Romeo LeBlanc, Minister of Public Works

MISSION: To provide property and operations management; administration of fire prevention and emergency preparedness, construction and realty technology development, and administration of technical expertise.

PRIMARY WORK: Directs government building construction programs; purchases and leases buildings; constructs and maintains roads and bridges; manages federal lands; performs building research to advance its mission.

SOURCE OF FINANCES: Parliament and government contracts.

PUBLICATIONS: Technical publications on a variety of subjects and annual listing of publications.

ONTARIO BUILDINGS BRANCH

777 Bay Street
2nd Floor
Toronto, Ontario M5G 2E5

CONTACT: David Hodgson, Director

MISSION: To develop and administer an effective regulatory system for the building industry in Ontario.

PRIMARY WORK: Develops and administers the Ontario Building Code, maintains the building industry regulatory system for Ontario, develops a code assessment framework for rational evaluation of existing and new code requirements, develops and implements educational courses for municipalities and industry, and computerizes building code and reference standards.

SOURCE OF FINANCES: Ontario government.

PUBLICATIONS: *Ontario Building Code, Building Action Newsletter, Guide and Illustrations for Building Code*, and *Building Industry Regulatory Reform in Ontario: A Proposal for a Building Code Assessment Framework*.

ONTARIO RESEARCH FOUNDATION

Sheridan Park Research Community
Mississauga, Ontario L5K 1B3

CONTACT: Lou Bruno, Manager, Energy Systems Centre

MISSION: To provide technical services to industry and government on a contract or fee basis.

PRIMARY WORK: Conducts research and development in building sciences, engineering, materials, and environment; assists clients in developing better products through product and process optimization, productivity improvements, and innovation.

SOURCE OF FINANCES: Research and development contracts.

PUBLICATIONS: Technical reports and specialty publications.

UNDERWRITERS' LABORATORIES OF CANADA (ULC)

7 Crouse Road
Scarborough, Ontario M1R 3A9

CONTACT: G. L. Toppin, Corporate Secretary

MISSION: To provide authoritative information to inspection authorities and others with regard to products, systems, and constructions having a bearing on fire, accident, or property hazards.

PRIMARY WORK: Certification, testing, and standards.

SOURCE OF FINANCES: Self-supporting.

PUBLICATIONS: ULC list of equipment and materials: Volume I "General" and Volume II "Building Construction."

UNIVERSITY OF MONTREAL, FACULTY OF DEVELOPMENT

BP 6128, Succ. A
Montreal, Quebec

CONTACT: Colin H. Davidson, Dean

MISSION: To advance building practices through research.

PRIMARY WORK: Performs research in building science and technology, the environment, urban planning, and man and the environment; offers education in architecture, landscape architecture, industrial design, and town planning.

SOURCE OF FINANCES: University.

PUBLICATIONS: Quarterly and biannual journals, research reports, and periodic reports.

CALGARY UNIVERSITY, DEPARTMENT OF CIVIL ENGINEERING

2500 University Drive, NW
Calgary, Alberta T2N 1N4

CONTACT: M. A. Sargious, Acting Head

MISSION: Education and research to support advances in technology.

PRIMARY WORK: Undertakes research in structures and solid mechanics, materials science, transportation engineering, water resources engineering, and geotechnical engineering.

SOURCE OF FINANCES: Province of Alberta, research grants, and contracts.

PUBLICATIONS: Research reports series, specialty publications on research results, and papers in referred journals.

NEW BRUNSWICK UNIVERSITY, DEPARTMENT OF CIVIL ENGINEERING (UNB/CE)

P.O. Box 4400
Fredericton, New Brunswick E3B 5A3

CONTACT: F. R. Wilson, Dean of Engineering

MISSION: Education and building research services.

PRIMARY WORK: Performs research on concrete, steel, geotechnical engineering, environmental and sanitary engineering, hydrology, and water engineering.

SOURCE OF FINANCES: Government grants, tuition fees, and research contracts.

PUBLICATIONS: Research reports and specialty reports.

WATERLOO UNIVERSITY, WATERLOO CONSTRUCTION COUNCIL (UW/WCC)

Department of Civil Engineering
University of Waterloo
Waterloo, Ontario N2l 3G1

CONTACT: R. W. Cockfield, Director

MISSION: To serve building industry organizations through research.

PRIMARY WORK: Provides research and training in construction management, methods, and equipment.

SOURCE OF FINANCES: Government and member dues.

PUBLICATIONS: Research reports, technical papers published in journals, and manuals.

UNIVERSITY OF WESTERN ONTARIO, BOUNDARY LAYER WIND TUNNEL LABORATORY (BLWT)

Faculty of Engineering Science
University of Western Ontario
London, Ontario N6A 5B9

CONTACT: A. G. Davenport, Director

MISSION: To further knowledge in wind engineering.

PRIMARY WORK: Performs research on wind engineering and structures, wind energy, environment, transportation, and wind and wave interaction.

SOURCE OF FINANCES: Research grants and contract research.

PUBLICATIONS: Research reports and specialty publications.

INTERNATIONAL DEVELOPMENT RESEARCH CENTRE (IDRC)

P.O. Box 8500
Ottawa, Ontario K1G 3H9

CONTACT: Aung Gyi, Associate Director, Cooperative Programs Division

MISSION: To stimulate and support scientific and technical research in developing countries.

PRIMARY WORK: Supports research in a variety of areas including industry, materials, and building technologies.

SOURCE OF FINANCES: Government.

PUBLICATIONS: Research reports.

CHILE

MINISTRY OF HOUSING AND URBANISM

Avda. Libertador B. O'Higgins
No. 924
Santiago

CONTACT: Miguel A. Poduje Sapiain, Minister of Housing and Urbanism
Luis Salas Romo, Subsecretary of Housing and Urbanism

MISSION: To improve the systems and building materials related to the construction of housing and equipment and to outline related guidelines and their legal implementation.

PRIMARY WORK: Disseminates the results of research and applies them to the contracts for housing and equipment that the ministry develops.

SOURCE OF FINANCES: Government.

PUBLICATIONS: Periodic bulletins and conference and seminar reports.

CHINA

CHINA ACADEMY OF BUILDING RESEARCH

9 Xiao Huang Zhuang
An Wai
Beijing

CONTACT: Zhengzhong Xu, President

MISSION: To serve as a comprehensive research institute.

PRIMARY WORK: Performs research and development on building technologies, including construction technology and products; compiles and administers building codes and standards; undertakes testing and measuring for quality control of construction engineering and products; provides testing services and technical training. Research is performed in the following areas: building structures, foundations engineering, earthquake engineering, building materials, building finishes, building physics, air conditioning, construction mechanization, comprehensive design of buildings, building fire, and computers.

SOURCE OF FINANCES: Partial government.

PUBLICATIONS: Bimonthly journal and specialty publications.

CHINA BUILDING TECHNOLOGY DEVELOPMENT CENTER

19 Che Gong Zhuang Street
Beijing

CONTACT: Ronglie Xu, Senior Engineer

MISSION: To promote the development of building technology and urban and rural construction.

PRIMARY WORK: Provides work in materials, renovation, assessment of building products, building economics, organization and dissemination of technology, training, compilation of building standards, and exhibitions.

SOURCE OF FINANCES: Government.

PUBLICATIONS: Bimonthly and quarterly publications, research reports, catalogues, journals, and periodic synopses.

SHANGHAI RESEARCH INSTITUTE OF BUILDING SCIENCES

75 Wan Ping Road
Shanghai

CONTACT: Pu Wang, Director

MISSION: To provide building research and testing services.

PRIMARY WORK: Performs research on structural engineering, building materials, construction machinery, building physics, computer techniques for construction, and economics and building information systems.

SOURCE OF FINANCES: Government ministries and research contracts.

PUBLICATIONS: Various publications on research activities and results.

TIANJIN FIRE RESEARCH INSTITUTE

P.O. Box 27
Tianjin

CONTACT: Bingyao Xu, Director

MISSION: To undertake research on fire growth and spread and applied techniques in fire prevention and extinguishment.

PRIMARY WORK: The Tianjin Fire Research Institute is involved in research on the combustion property of materials and techniques in fire prevention and explosion venting; the theory of fire modeling and testing techniques; techniques of fire statistics, fire investigation, and fire-cause determination; fire extinguishants, fire resistance, and fire-resistant materials; applied techniques of automatic fire alarm and automatic fire-extinguishing engineering; fire standards and fire prevention codes. The institute is also the National Center for Inspection and Testing of Product Quality of Fixed Fire Extinguishing System and Fire Resistant Building Components. The center tests fire extinguishants, fire-extinguishing systems' components, and building components.

SOURCE OF FINANCES: Government and fees for services.

PUBLICATIONS: Journal and newsletter.

COLOMBIA

COLOMBIAN INSTITUTE OF CEMENT PRODUCERS

P.O. Box 52816
Medellin

CONTACT: Carlos A. Ossa, Executive Director

MISSION: To undertake technical activities for the better use of cement.

PRIMARY WORK: Coordinates programs to raise the level of technical processes in use in the cement factories.

SOURCE OF FINANCES: Industry.

PUBLICATIONS: Technical reports and bulletins.

CZECHOSLOVAKIA

BUILDING INSTITUTE FOR TESTING MATERIALS AND STRUCTURES (TZUS)

Konviktska 24
113 33 Praha 1

CONTACT: Jiri Petran, Managing Director

MISSION: Research and testing of materials and structures for improved building practices.

PRIMARY WORK: Performs research on structures and materials, develops and verifies test procedures and test apparatus, and encourages state building inspections.

SOURCE OF FINANCES: Government and contract research.

PUBLICATIONS: Quarterly and monthly bulletins of test procedures and research results; annual reports on defects in structures.

BUILDING RESEARCH INSTITUTE (VUPS)

102 21 Praha 10
Hostivar
Prazska 16

CONTACT: Josef Sovadina, Director

MISSION: To focus on the technical development and reconstruction of residential, public, industrial, and agricultural structures.

PRIMARY WORK: Performs research on the theory of building structures in the field of building physics and chemistry including fire safety; research on the technical development of residential, public, industrial, and agricultural structures, as well as their elements, components, and whole systems; new building technologies; and research on the reconstruction and modernization of residential, public, and industrial buildings.

SOURCE OF FINANCES: Government and building industry enterprises contracts.

PUBLICATIONS: Publications of research results and technical reports for contractors.

CZECHOSLOVAK BUILDING CENTRE

Vaclavske namesti 31
111 21 Praha 1

CONTACT: Miloslav Matasovsky, Director

MISSION: To influence the development and quality of construction materials and products.

PRIMARY WORK: Provides services to advance information systems; develops methods to assess materials, performs the testing of quality of building products, and serves as government testing laboratory.

SOURCE OF FINANCES: Fees for information services, government, and contracts.

PUBLICATIONS: Quarterly, biannual, and annual publications on building technology; catalogue of materials; and manuals.

RESEARCH INSTITUTE FOR BUILDING AND ARCHITECTURE (VUVA)

Letenska 3
118 00 Praha 1
Mala Strana

CONTACT: Vaclav Chyska, Director

MISSION: To serve as central source for building research services.

PRIMARY WORK: Performs research in areas of housing, architecture, urbanization, town-planning, urban planning, economics, environment, and experimental design; disseminates results to building community; and conducts training conferences.

SOURCE OF FINANCES: Government and parent organizations.

PUBLICATION: Monthly journal.

RESEARCH CENTER FOR CIVIL ENGINEERING (VUIS)

815 37 Bratislava
Lamacske 8

CONTACT: L. Kompis

MISSION: To perform civil engineering research and documentation services for the building industry.

PRIMARY WORK: Provides research and consulting services in areas such as bridges, roads, hydraulic construction, tunnels and other underground structures, and soil and rock mechanics; licenses foreign technologies.

SOURCE OF FINANCES: Government and contracted services.

PUBLICATIONS: Research reports, quarterly bulletin, and annual reviews of activities.

**SLOVAK ACADEMY OF SCIENCES,
INSTITUTE FOR CONSTRUCTION AND ARCHITECTURE (USTARCH SAV)**

Dubravska cesta
842 20 Bratislava

CONTACT: Rudolf Skrucany, Director

MISSION: To serve as a central institute for complex basic and applied building research.

PRIMARY WORK: Performs research on complex structural problems, nonlinear and fracture mechanics, failure of reinforced concrete and metalic structures, building materials, building physics, urban and rural planning, and influence of social factors.

SOURCE OF FINANCES: Slovak Academy of Sciences.

PUBLICATIONS: Research reports and periodicals.

DENMARK

CEMENT AND CONCRETE LABORATORY

Aalborg Portland
Rordalsveg 44
P.O. Box 165
DK 9100

CONTACT: Palle Nepper-Christensen, Manager

MISSION: Research, development, and control for better production and use of cement.

PRIMARY WORK: Provides technical and laboratory support for improving quality of cement, concrete, refractories, and fiber-reinforced composites.

SOURCES OF FINANCE: Aktiesselskabet Aalborg Portland-Cement-Fabrik.

PUBLICATIONS: Research reports.

CONCRETE AND STRUCTURAL RESEARCH INSTITUTE (BKI)

Dr. Neergaards Vej 13
DK-2970 Horsholm

CONTACT: S. Oivind Olesen

MISSION: To provide testing and research in concrete and structural technology.

PRIMARY WORK: Performs research on structures and structural members, concrete materials, material deterioration, computer program selection, and admixtures for concrete; conducts educational programs; develops and sells new computer programs such as MISTRA, a comprehensive, interactive program for structural design calculations.

SOURCES OF FINANCE: Contract income; public and private funds.

DANISH BUILDING CENTER

Vester Voldgade 94
DK 1552 Copenhagen V

CONTACT: Jorgen Bagge Andersen, President

MISSION: To serve the building community with information, testing services, and documentation.

PRIMARY WORK: Provides general information about solving building problems through consulting, a documentation center, permanent exhibit, and training courses.

SOURCE OF FINANCES: Sales of information, training fees, and consulting services.

PUBLICATIONS: Manuals, failure sheets, legislation, standards, and special purpose reports.

DANISH BUILDING RESEARCH INSTITUTE (SBI)

P.O. Box 119
DK 2970 Horsholm

CONTACT: Hans Jorgen Larsen, Director

MISSION: National building research laboratory focusing on improving building practices.

PRIMARY WORK: Provides research in areas such as acoustics, building physics, structural engineering, building and environmental design, urban and regional planning, and building economics; has an acoustics testing laboratory.

SOURCES OF FINANCES: Mainly government.

PUBLICATIONS: Research reports, bulletins, and articles in professional journals.

DANISH CORROSION CENTER

Park Alle 345
DK-2605 Brondby

CONTACT: Hans Arup, Director

MISSION: To provide technical assistance on the physical and chemical behavior of metals.

PRIMARY WORK: Performs research and testing on metals in concrete to assess corrosion of reinforcements, metals in water systems and heating systems, metallic coatings, and paints as corrosion protection.

SOURCE OF FINANCE: Government and grants.

PUBLICATIONS: Research reports and newsletter.

DANISH FIRE PROTECTION ASSOCIATION

Datavej 48
DK-3460 Birerod

CONTACT: Ernst Nielson, Managing Director

MISSION: To conduct research to prevent fire and to limit the economic and human consequences of fire.

PRIMARY WORK: Provides fire safety inspections, fire site investigations, education and training programs, research and development activities, and the dissemination of information to the public.

SOURCE OF FINANCES: Danish Council of Technology, Ministry of Industry (10%); paid activities (90%).

PUBLICATIONS: Guidelines.

DANISH ILLUMINATION ENGINEERING LABORATORY (LTL)

Bygning 325
Lundtoftevej 100
2800 Lyngby

CONTACT: Bjarne Nielsen, Director

MISSION: Research and testing in the area of illuminating engineering.

PRIMARY WORK: Research, testing, and development of products and computer programming related to work lighting, room lighting, reflective properties of road surfaces, and road markings.

SOURCE OF FINANCES: Government and contract research.

PUBLICATIONS: Journal, research notes, and technical reports.

DANISH LIME AND BRICK LABORATORY

Teglbakvej 20
8361 Hasselager

CONTACT: E. Kjaer, Director

MISSION: To provide research and testing for manufacturers and users of bricks.

PRIMARY WORK: Performs physical and chemical tests and analyses of raw materials used in brick manufacturing and use.

SOURCE OF FINANCES: Contract research, brick and lime association, and government.

PUBLICATIONS: Laboratory and conference reports.

DANISH NATIONAL TESTING BOARD

The Danish National Agency of Technology
Secretariat of the National Testing Board
Tagensvej 135

CONTACT: H. H. Bernth, Head of Secretariat for National Testing Board

MISSION: To coordinate the technical testing facilities within Denmark and to administer an accreditation scheme for testing and calibration laboratories.

PRIMARY WORK: Coordinates Danish testing resources and administers an accreditation scheme that comprises a large number of Danish laboratories which have been accredited by the board to carry out technical testing.

SOURCE OF FINANCES: Government; fees for accreditation.

PUBLICATIONS: Requirements; guidelines; register of accredited laboratories; general information on the accredition scheme.

DANTEST-NATIONAL INSTITUTE FOR TESTING AND VERIFICATION

Amager Boulevard 115
DK 2300 Copenhagen S

CONTACT: E. Repstorff Hotlveg, Director

MISSION: To serve as principal testing laboratory and metrology institute.

PRIMARY WORK: Provides services in areas such as fire technology, mechanics, structural engineering, chemical technology, calibration and verification of equipment, and product control procedures; develops new test procedures; performs technical advisory services.

SOURCES OF FINANCES: Grants and fees for services.

PUBLICATIONS: Technical standards, reports, monographs, and brochures.

**THE ROYAL DANISH ACADEMY OF ART,
SCHOOL OF ARCHITECTURE**

Kongens Nytorv 1
DK 1050 Copenhagen K

CONTACT: Ebbe Harder, Registrar

MISSION: To provide teaching and laboratory facilities.

PRIMARY WORK: Conducts research in architectural history, building, building restoration, design, and rural and urban planning; provides special laboratory facilities for daylighting observation and full-scale modeling.

SOURCES OF FINANCE: State and fund financed.

PUBLICATIONS: Internal research newsletter.

**ROYAL DANISH ACADEMY OF FINE ARTS,
SCHOOL OF ARCHITECTURE, INSTITUTE OF BUILDING SCIENCE,
DEPARTMENT OF BUILDING SCIENCE**

Peder Skramsgade 1
DK 1054 Copenhagen K

CONTACT: Bjorn Bindsley, Technical Director

MISSION: To provide services to advance building practices.

PRIMARY WORK: Performs services through education and technical assistance in planning, documenting, and managing building projects; undertakes research and devlopment of computer-aided project management systems.

SOURCE OF FINANCES: Government

PUBLICATION: Proceedings and course work.

**SOFUS-BYG--COOPERATION ON RESEARCH, DEVELOPMENT, AND
TECHNOLOGICAL SERVICES IN THE BUILDING SECTOR**

c/o Danish Building Research Institute
Postbox 119
DK-2970 Horsholm

MISSION: To provide advice and guidance to the 23 member organizations on topics of building technology.

PRIMARY WORK: Provides advice, mediation, and coordination of all research organizations in the cooperative; organizes symposias and work groups to address relevant technological questions, e.g., fire technology, energy research and development, and load-bearing construction.

PUBLICATION: Catalogue.

TECHNICAL UNIVERSITY OF DENMARK

DK 2800 Lyngby

CONTACT: Torben C. Hansen, Director, Building Materials Laboratory

MISSION: Education, research, and testing to advance test and measurement methods, building standards, and building practices.

PRIMARY WORK: Performs work in the following areas: building materials; structural reliability and design; soil and rock mechanics; foundation engineering; environmental engineering; hydrology and coastal engineering; building physics; heat production, distribution, and recovery; building technology; planning and executing construction contracts; and managerial related assistance.

SOURCES OF FINANCE: Government, Commission of the European Communities, and sponsored research.

PUBLICATIONS: Professional and scientific journals, theses, research reports, bulletins, periodic reports, and annual lists of publications.

ECUADOR

GUAYAQUIL CHAMBER OF CONSTRUCTION

Simon Rodriguez y Pedro Gaul
CDLA Bolivariana
P.O. Box 8955
Guayaquil

CONTACT: Hernan Benites, President
Antonio Belrran, General Manager

MISSION: To further the development of the construction industry and the search for better conditions for construction activities.

PRIMARY WORK: Provides direct services to associates; analyzes economic trends as they affect the construction industry; proposes laws and by-laws for the improvement of the legal conditions for the construction industry; and protects contractors under difficult operating conditions.

SOURCE OF FINANCES: Membership fees and related services.

PUBLICATIONS: Bimonthly technical bulletins.

EGYPT

GENERAL ORGANIZATION FOR HOUSING, BUILDING, AND PLANNING RESEARCH (GOHBP)

El-Tahreer Street
Dokki, Giza
P.O.Box 1770
Cairo

CONTACT: Abov Zeid Ragrh, Chairman

MISSION: To provide research and technical development in the fields of housing, building, and planning.

PRIMARY WORK: Performs work in foundations and underground structures, regional and town planning, building materials, building physics, and in financial and administrative management.

SOURCE OF FINANCES: Government, contract research, and international grants.

PUBLICATIONS: Professional journals and research reports.

FEDERAL REPUBLIC OF GERMANY

FRAUNHOFER INSTITUTE FOR WOOD RESEARCH, WILHELM KLAUDITZ INSTITUT

Bienroder Weg 54 E
D-3300 Braunschweig

CONTACT: G. Kossatz, Director

MISSION: To improve wood as a raw material; to develop wood-based materials; to investigate wood's practical application; and to undertake research on adhesives and surface coatings for wood and wood-based products, process automation, properties and use of wood-based building materials and structural members, and investigations on environmental pollution (analyses of emissions).

PRIMARY WORK: In addition to research and development, seminars and conferences on wood and wood products are held for professionals and other interested persons.

SOURCE OF FINANCES: Government and contract research financed by industry and business associations.

PUBLICATIONS: Journals; research reports.

FEDERAL INSTITUTE FOR MATERIALS RESEARCH AND TESTING (BAM)

Unter den Eichen 87
D 1000 Berlin 45

CONTACT: G. W. Becker, President

MISSION: To establish test methods and perform materials research and chemical safety engineering.

PRIMARY WORK: Performs work in developing test methods and testing building materials; training; and performs research in metals, structures, organic materials, materials nondestructive testing, and chemical engineering safety.

SOURCE OF FINANCES: Government and fees for research.

PUBLICATIONS: Quarterly journal, annual report, technical standards, research reports, and documentation.

FEDERAL HIGHWAY RESEARCH INSTITUTE (BAST)

Bruederstrasse 53
D 5060 Bergisch Gladbach 1

CONTACT: Heinrich Praxenthaler, President

MISSION: To serve as central research institute for highway design and construction.

PRIMARY WORK: Provides services in highway construction and networks, environmental impact, and traffic safety.

SOURCE OF FINANCES: Government.

PUBLICATIONS: Serial publications, papers, and special reports.

FEDERAL WATERWAY ENGINEERING AND RESEARCH INSTITUTE (BAW)

Kussmaulstrasse 17
D-7500 Karlsruhe 21

CONTACT: H. P. Tzschucke, Head of the Construction Department

MISSION: The Department of Civil Engineering Construction within the BAW serves as a research and advisory center for the Federal Waterway Authorities.

PRIMARY WORK: Analysis and expert opinion about existing civil engineering construction and materials; research and advice for the repair of old construction; and collection and evaluation of practical experiences in the fields of concrete and steel structures, corrosion protection, and materials for civil engineering.

SOURCE OF FINANCES: Government.

PUBLICATIONS: "Mitteilungsblatt der BAW"; "BAW-Brief."

GERMAN SOCIETY OF MASONRY CONSTRUCTION

Kortumstrasse 50
4300 Essen

CONTACT: Klaus Gobel, President

MISSION: To serve as the technical laboratory for masonry producers.

PRIMARY WORK: Technical assistance to improve masonry problems.

SOURCE OF FINANCES: Membership fees.

PUBLICATIONS: Technical standards.

RESEARCH ASSOCIATION FOR BUILDING AND HOUSING

Silberbugstrasse 160
7000 Stuttgart 1

CONTACT: J. Herkommer, Chairman of Board

MISSION: To serve as the central source for providing technical assistance to the building community.

PRIMARY WORK: Performs research on building, housing economics, and town and regional planning.

SOURCE OF FINANCES: Government.

PUBLICATIONS: Research reports and specialty publications.

RESEARCH INSTITUTE OF THE CEMENT INDUSTRY, ASSOCIATION OF THE GERMAN CEMENT WORKS (VDZ)

Tannenstrasse 2
Postfach 30 10 63
D 4000 Dusseldorf 30

CONTACT: J. Bonzel and F. W. Locher, Heads

MISSION: To serve as a technical resource for the manufacturers and users of cement.

PRIMARY WORK: Performs research in cement and concrete technology, chemistry, manufacturing, environmental protection, and safety.

SOURCE OF FINANCES: Membership dues.

PUBLICATIONS: Research reports, biannual reports, information circulars, and technical notes.

INSTITUTION FOR RESEARCH AND MATERIAL TESTING (FMPA)

Plaffenwaldring 4
D 7000 Stuttgart 80

CONTACT: Gallus Rehm, Director

MISSION: To serve as a research facility for testing and development.

PRIMARY WORK: Performs research on building materials and structures, fires, and performs chemical analysis.

SOURCE OF FINANCES: Government grants and fees for research.

PUBLICATIONS: Research reports, bibliographies, and professional and scientific journals.

FRAUNHOFER INSTITUTE FOR BUILDING PHYSICS (IBP)

Nobelstrasse 12
D 7000 Stuttgart 80 (Vaihingen)

CONTACT: Fridolin P. Mechel, Director, Acoustics Department
Karl Gertis, Director, Heat/Climates Department

MISSION: To serve as primary research and testing laboratory.

PRIMARY WORK: Performs research in architectural and industrial acoustics, heat and moisture migration, energy conservation, indoor climate, and weathering; develops test methods for construction and structural elements.

SOURCE OF FINANCES: Fees for research and government grants.

PUBLICATIONS: Research results and specialty publications.

INFORMATION CENTER FOR REGIONAL PLANNING AND CONSTRUCTION (IRB)

Nobelstrasse 12
D 7000 Stuttgart 80

CONTACT: W. Wissmann, Director

MISSION: To serve as the central information agency for building construction, town planning, and regional planning and housing.

PRIMARY WORK: Collects and analyzes technical and scientific information from literature and research sources; stores information on a data base from which all information services are derived.

SOURCE OF FINANCES: Government and sales of services.

PUBLICATIONS: Bibliographies, literature information services, specialized periodicals with references to new publications and short reports on building research, catalogues of building research reports, FINDEX (facet-oriented indexing system), and ICONDA Communication Format (exchange format for the International Construction Data Base).

INSTITUTE OF BUILDING RESEARCH (IFB)

An der Markuskirche 1
D 3000 Hannover 1

CONTACT: Herbert Menkhoff, Director

MISSION: To serve as the central source for building research.

PRIMARY WORK: Performs research in physics and construction, techniques and experimental construction, and planning and productivity.

SOURCE OF FINANCES: Fees for research, grants, and government.

PUBLICATIONS: Monthly journal, research reports, professional journal, and specialized publications.

MATERIALS TESTING INSTITUTE, NORTH-RHINE-WESTPHALIA

P.O. Box 41 03 07
Marsbruchstr 186
D-4600 Dortmund 41

CONTACT: A. Kremeier, Director

MISSION: To serve as a test institute for industry and public authorities.

PRIMARY WORK: Tests materials and structures; conducts applied research and production conformity control.

SOURCE OF FINANCES: Government.

PUBLICATIONS: Biannual reports and special publications of staff members in periodicals.

RESEARCH ASSOCIATION FOR UNDERGROUND TRANSPORTATION FACILITIES, INC. (STUVA)

Mathias-Bruggen-Str. 41
D-5000 Koln 30

CONTACT: A. Haack, Manager

MISSION: Independent research institute in the fields of urban traffic (railway, road, pedestrian traffic) in tunnels and on the surface, environmental protection (noise, vibration, fumes), tunneling technology, and underground services and utility tunnels.

PRIMARY WORK: Research and consulting, testing, information, and data bases.

SOURCE OF FINANCES: Government and contracts.

PUBLICATIONS: Journal, book series, and research reports.

INSTITUTE OF CIVIL ENGINEERING AND BUILDING TECHNOLOGY (IFBT)

Reichpietschufer 72-76
1000 Berlin 30

CONTACT: G. Breitschaft, President

MISSION: To serve as a central source for civil engineering and building technology issues.

PRIMARY WORK: Provides services in civil engineering, prestressed steel, metallurgy, public works, concrete, plastics, building physics, fire protection, documentation, standardization; issues approval marks on quality control.

SOURCE OF FINANCES: Government and fees for approvals.

PUBLICATIONS: Bimonthly periodicals and lists of approvals.

INSTITUTE FOR REINFORCED CONCRETE (IBS)

Landsbergerstrasse 414
8000 Munchen 60

CONTACT: H. Martin, Director

MISSION: To serve members with laboratory and testing expertise.

PRIMARY WORK: Performs research and testing in reinforced concrete structures, methods of analysis, durability, and test methods.

SOURCE OF FINANCES: Contract research and member fees.

PUBLICATIONS: Research reports and professional journals.

BUILDING DIVISION OF THE GERMAN STANDARDS INSTITUTION

Burggrafenstrasse 4-10
Postfach 1107
D 1000 Berlin 30

CONTACT: H. Bub, President

MISSION: To establish and coordinate all building standards and codes.

PRIMARY WORK: Through more than 300 working committees develops basic and planning standards, measuring standards, unified technical building regulations; addresses town and country planning, foundations, building components and elements, concrete, steel construction, finishings, roads, highways and bridges, and industrial and agricultural buildings.

SOURCE OF FINANCES: Government and associations interested in standardization.

PUBLICATIONS: Periodic issues of standardization news and technical standards and regulations.

COMPUTING AND DEVELOPMENT CENTER FOR DATA PROCESSING IN CIVIL ENGINEERING (RIB)

Schulze-Delitzsch-Strasse 28
D 7000 Stuttgart 80

CONTACT: Wolfgang Haas, General Manager

MISSION: Civil engineering research and software development.

PRIMARY WORK: Provides services in development of computer-aided design, computer-integrated drafting, structural analysis, highway design, and quantity surveying.

SOURCE OF FINANCES: Contracted research and sale of software.

PUBLICATIONS: Research reports, technical papers, and user manuals.

AACHEN TECHNICAL UNIVERSITY, INSTITUTE FOR BUILDING RESEARCH

Schinkelstrasse 3
D-5100 Aachen

CONTACT: H. R. Sasse, Director
P. Schiebl, Director
P. Schubert

MISSION: Research, testing, development, and application of building materials.

PRIMARY WORK: Research and development in the field of building materials, especially concrete, steel, masonry, use of waste materials, and polymers; special emphasis on basic behavior and properties, testing methods, and durability.

SOURCE OF FINANCES: Government and contract research.

PUBLICATIONS: Annual report, publications, and books.

**DARMSTADT TECHNICAL UNIVERSITY,
INSTITUTE FOR CONCRETE STRUCTURES (THD/IFM)**

Alexanderstrasse 5
6100 Darmstadt

CONTACT: Gert Konig, Director IFM
Joost Walraven
Hans-Wolf Reinhardt

MISSION: To serve as technical laboratory resource to the building community.

PRIMARY WORK: Performs research in structural and earthquake analysis, materials, reinforced concrete, and prestressed concrete detailing and design.

SOURCES OF FINANCE: Government, research contracts, and university.

PUBLICATIONS: Research reports, papers published in professional journals, and *Darmstadt Concrete.*

**DARMSTADT TECHNICAL UNIVERSITY,
INFORMATION PROCESSING IN BUILDING AND CIVIL ENGINEERING
ORGANIZATIONS (IVB)**

Petersenstrasse 13
61000 Darmstadt

CONTACT: Heinz Schwarz, Director

MISSION: To serve as technical laboratory resource to the building community.

PRIMARY WORK: Performs work in building planning and design, information storage and retrieval, and evaluation of software and test procedures.

SOURCE OF FINANCES: Government, research contracts, and university.

PUBLICATIONS: Research reports, papers published in professional journals, and specialty publications.

**DARMSTADT TECHNICAL UNIVERSITY,
INSTITUTE OF STEEL CONSTRUCTION**

Technical University
Alexanderstrasse 7
61000 Darmstadt

CONTACT: J. G. Bouwkamp

MISSION: Research on design, fabrication, and construction of land-based and offshore steel structures; materials testing; proof testing under static, cyclic, and random dynamic loads; service to industry; and continuing education.

PRIMARY WORK: Static, cyclic, dynamic, and pseudodynamic testing of piping elements, systems, and support devices; earthquake resistance of piping systems and composite steel/concrete connections, assemblies, and frames under pseudodynamic loads using integrated CAT (Computer-Aided Testing) methods; full-scale, on-site dynamic studies of buildings, bridges, towers, and other civil engineering structures through field measurements; fire testing of connections and members; advisory services on structural engineering, including linear and nonlinear stability; and development of CAL (Computer-Aided Learning) and CAE (Computer-Aided Engineering) systems for teaching and design.

SOURCE OF FINANCES: Government and service to industry.

PUBLICATIONS: Research reports and dissertations.

**BERLIN TECHNICAL UNIVERSITY,
INSTITUTE OF SOIL MECHANICS AND FOUNDATION ENGINEERING**

Strasse des 17. Juni 135, Sekr. B7
D-1000 Berlin 12

CONTACT: S. Savidis, Director

MISSION: Education and research work in civil engineering and consulting.

PRIMARY WORK: Lectures and seminars in soil mechanics, dynamics, and foundation engineering; theoretical and experimental research activities, e.g., in earthquake engineering with numerical investigations and model tests on a shake table; consulting and laboratory investigations.

SOURCE OF FINANCES: Government, research projects, consulting, and sales of services (laboratory).

PUBLICATIONS: Research reports.

**BERLIN TECHNICAL UNIVERSITY,
INSTITUTE FOR STRUCTURAL DESIGN AND MATERIAL STRENGTH (IBF),
HERMANN RIETSCHEL INSTITUTE FOR HEATING AND AIR-CONDITIONING
ENGINEERING (HRI)**

Strasse des 17 Juni 135
D 1000 Berlin 12

CONTACT: M. Specht, Director, IBF
Horst Esdom, Managing Director, HRI

MISSION: To provide teaching and research laboratory capabilities.

PRIMARY WORK: IBF performs research on buildings and bridges, prefabricated construction, light metal construction, and concrete materials. HRI performs research on fundamentals of HVAC and energy conservation and tests heating components.

SOURCE OF FINANCES: University, industry, and contracted research.

PUBLICATIONS: HVAC text books, research results, and professional journals.

BRAUNSCHWEIG TECHNICAL UNIVERSITY, INSTITUTE OF BUILDING MATERIALS, REINFORCED CONCRETE AND FIRE PROTECTION

Beethovenstrasse 52
3300 Braunschweig

CONTACT: Karl Kordina and F. S. Rostasy, Directors

MISSION: To educate and serve as a research laboratory.

PRIMARY WORK: Performs research and testing of building materials and reinforced and prestressed structural elements, acoustics, chemistry and physics of building materials, fire testing, behavior of materials in fire, and fire engineering design.

SOURCE OF FINANCES: Government, research contracts, and fees for testing.

PUBLICATIONS: Research reports and publications in professional journals.

BRAUNSCHWEIG TECHNICAL UNIVERSITY, INSTITUTE FOR STATICS (STRUCTURAL ANALYSIS)

Beethovenstrasse 51
D-3300 Braunschweig

CONTACT: Heinz Duddeck

MISSION: Research institute for analysis of civil engineering and underground structures.

PRIMARY WORK: Analysis in structural engineering, tunnels, nonlinear methods, plasticity, and underground structures in rock and salt rock.

SOURCE OF FINANCES: Government and research societies.

PUBLICATIONS: Reports on research work and periodicals.

ESSEN UNIVERSITY, INSTITUTE OF BUILDING PHYSICS

P.O. Box 68 43
D 4300 Essen 1

MISSION: To provide academic training and to serve as research laboratory.

PRIMARY WORK: Performs work in energy conservation, thermal insulation, solar technology, acoustics, moisture migration, thermal stresses on structures, and environmental issues.

SOURCE OF FINANCES: Research contracts and university.

PUBLICATIONS: Research reports and specialty reports.

HANNOVER UNIVERSITY, INSTITUTE OF BUILDING MATERIALS AND MATERIALS TESTING

Nienburger Strasse 3
D-3000 Hannover 1

CONTACT: H. J. Wierig

MISSION: Lectures and research in the field of building materials, development of technologies, and quality control.

PRIMARY WORK: Research on concrete, masony, and mortars.

SOURCE OF FINANCES: Government, governmental and industrial research funds, and sales of services.

PUBLICATIONS: *Mitteilungen aus dem Institute fur Baustoffkunde und Materialprufung der Universitat Hannover.*

UNIVERSITY OF HANNOVER, INSTITUTE FOR HYDRAULICS AND COASTAL ENGINEERING

Nienburger Strasse 4
D-3000 Hannover

CONTACT: Managing Director

MISSION: Execution of scientific studies in hydraulics and coastal engineering; special interest in waves, tides and sedimentation, harbours, estuaries and coastal protection, hydraulic structures, hydraulic modeling.

SOURCE OF FINANCES: Governmental institutions, research funds, and clients in Germany and abroad.

PUBLICATION: *Mitteilungen des Franzius-Instituts* (biannually).

STUTTGART UNIVERSITY, INSTITUTE FOR CONCRETE STRUCTURES

Pfaffenwaldring 7
D-7000 Stuttgart 80

CONTACT: E. H. J. Schlaich, Director

MISSION: Research institute serving the University of Stuttgart.

PRIMARY WORK: Teaching and theoretical and experimental research in the field of reinforced and prestressed concrete structures; code work in cooperation with national and international organizations; expertises and research in the field of cable net structures and mixed-use structures.

SOURCE OF FINANCES: Government, research funds of organizations, and industry.

PUBLICATIONS: Research reports, papers in technical journals, and textbooks.

STUTTGART UNIVERSITY, INSTITUTE OF CONSTRUCTION MANAGEMENT, INSTITUTE OF STRUCTURAL ANALYSIS

Pfaffenwaldring 7
Postfach 80 1140
D 7000 Stuttgart 80

CONTACT: G. Drees, Director
E. Ramm, Director

MISSION: To provide academic education and serve as a research laboratory.

PRIMARY WORK: Performs research in construction management, organization of construction, energy use, cost control and structural analysis, computer programming, stability of structures, structural optimization, project management, and organization of construction companies, and controlling.

SOURCES OF FINANCE: Government.

PUBLICATIONS: Research reports, bulletins, technical and professional journals, and textbooks.

STUTTGART UNIVERSITY, GEOTECHNICAL INSTITUTE FOR UNDERGROUND BUILDING, SOIL MECHANICS, ROCK MECHANICS, AND TUNNEL CONSTRUCTION

Pfaffenwaldring 35
P.O. Box 80 11 40
D-7000 Stuttgart 80

CONTACT: U. Smoltczyk

MISSION: Research laboratory for soil mechanics, rock mechanics, foundation engineering, and tunneling problems.

PRIMARY WORK: Research, teaching, and consulting.

SOURCE OF FINANCES: Govermental bonds and sales of services.

PUBLICATIONS: Research reports.

**UNIVERSITY OF KARLSRUHE,
CHAIR OF SOIL MECHANICS AND FOUNDATION ENGINEERING,
INSTITUTE OF SOIL MECHANICS AND ROCK MECHANICS**

Kaiserstrasse 12
D-7500 Karlsruhe 1

CONTACT: D. Kolymbas, Chief Engineer

MISSION: University teaching, research, and consulting.

PRIMARY WORK: Teaching, fundamental and applied research on soil behavior, numerical techniques, stability analysis of foundations and earth constructions, model tests (tunnels, retaining walls, reinforced soil), frozen soil laboratory, soil dynamics, and plastodynamics.

SOURCES OF FINANCES: Government and sales of services.

PUBLICATIONS: Conference reports and journals.

**KARLSHRUHE UNIVERSITY,
DEPARTMENT OF STEEL AND ALUMINUM CONSTRUCTION,
BUILDING MATERIALS TESTING INSTITUTE**

Kaiserstrasse 12
D 7500 Karlshruhe 1

CONTACT: H. K. Hilsdorf, Director

MISSION: Education, materials research, surveillance, quality control, and proof testing.

PRIMARY WORK: Mechanical properties of building materials, especially concrete; durability of building materials and members; development of new building materials; microstructure of materials; and reinforced masonry.

SOURCE OF FINANCES: Government, National Science Foundation, and industry.

PUBLICATIONS: Research results, related contributions in various national and international scientific magazines, conference proceedings, and books.

FINLAND

ASSOCIATION OF FINNISH CIVIL ENGINEERS (RIL)

Meritullinkatu 16A
00170 Helsinki 17

CONTACT: Yrjo Matikainen, Managing Director

MISSION: To improve the professional practice of Finnish civil engineers.

PRIMARY WORK: Develops improved civil engineering practices and quality control, performs research in civil engineering, contributes to improving building standards, conducts postgraduate courses, and holds design competitions.

SOURCE OF FINANCES: Membership dues, consulting fees, and sales of documents.

PUBLICATIONS: Journal, technical standards, handbooks, training aids, and catalogues.

BRICK LABORATORY OF THE FINNISH BRICK INDUSTRY ASSOCIATION

Laturinkuja 2
Box 6
SF-02601 Espoo

CONTACT: Martti Romu, Laboratory Director

MISSION: To provide technical services to the Finnish brick industry.

PRIMARY WORK: Performs testing and research on brick and clay products.

SOURCE OF FINANCES: Sales of services.

PUBLICATIONS: Technical recommendations for masonry materials.

BUILDING INFORMATION INSTITUTE

Lonnrotinkatu 20 B
SF 00120 Helsinki 12

CONTACT: Esko Lehti, Director General

MISSION: To promote information services, research, and standardization of design and construction practices.

PRIMARY WORK: Performs work leading to standardization of building design and construction practices, develops standards and regulations, and disseminates information to the building community through product files and information files.

SOURCE OF FINANCES: Sales of publications, fees for services, and government.

PUBLICATIONS: Various files, product index, quarterly reports, and specialty publications.

CONCRETE ASSOCIATION OF FINLAND

Mikonkatu 18 B 12
SF 00100 Helsinki 10

CONTACT: Heikki Kaitila, Managing Director

MISSION: To promote the development of concrete technology, structural design, and production of concrete structures through technical and scientific work.

PRIMARY WORK: Undertakes work in standardization, training, and research.

SOURCE OF FINANCES: Sales of documents and membership fees.

PUBLICATIONS: Journal, standards and codes of practices, research reports, annual report, and training publications.

HELSINKI UNIVERSITY OF TECHNOLOGY, FACULTY OF SURVEYING AND CIVIL ENGINEERING

Rakentajanaukio 4 A
02150 Espoo 15

CONTACT: M. Mikkola, Dean

MISSION: Education and research on civil engineering and surveying problems.

PRIMARY WORK: Performs work in in the following areas of building technology: structural mechanics; bridge construction; soil mechanics; transportation and highway engineering; water resources; hydraulic and sanitary engineering; building economics; and photogrammetry.

SOURCE OF FINANCES: Government and research contracts.

PUBLICATIONS: Research reports and other specialty publications.

TAMPER UNIVERSITY OF TECHNOLOGY, DEPARTMENT OF CIVIL ENGINEERING

P.O. Box 527
SF 33101 Tampere 10

CONTACT: Olli-Pekka Hartikainen, Director

MISSION: To educate and perform research to improve civil engineering practices.

PRIMARY WORK: Provides research in the following areas: earth and highway construction, engineering geology, soil mechanics, structural loads, photogrammetry, building construction, and economics.

SOURCE OF FINANCES: University and research contracts.

PUBLICATIONS: Results of research and other specialty publications.

**FINLAND TECHNICAL RESEARCH CENTRE,
DIVISION FOR BUILDING TECHNOLOGY AND COMMUNITY
DEVELOPMENT (VTT)**

Vuoimiehentie 5
02150 Espoo 15

CONTACT: P. Jauho, General Director

MISSION: National research laboratory and testing institute.

PRIMARY WORK: Performs research in the following areas: soils and foundations, structural mechanics, concretes, fire technology, heating and ventilating, land use, renovation, forest products related to raw materials, economics, community planning, environmental protection, road construction and traffic engineering.

SOURCE OF FINANCES: Government and fees for services.

PUBLICATIONS: Research reports and notes, proceedings, and annual report.

FRANCE

FRENCH PRECAST CONCRETE STUDY AND RESEARCH CENTER (CERIB)

rue des Longs Reages
BP 59
F 28230 Epernon

CONTACT: Michel Darcemont, General Director

MISSION: To serve as the technical industrial center for French precast concrete.

PRIMARY WORK: Undertakes research dealing with the manufacture and use of concrete products, provides technical assistance, disseminates information, trains engineers and technicians, performs quality control and standardization through AFNOR (French quality mark), and prepares codes of practice.

SOURCE OF FINANCES: Levy on concrete manufacturers and contracts for research.

PUBLICATIONS: Technical publications, monographs, and quarterly reports.

**MEDITERRANEAN CENTER FOR RESEARCH AND APPLIED STUDIES IN
INDUSTRY AND CONSTRUCTION (CEMEREX)**

rue no. 14
13127 Vitrolles
Marseille

CONTACT: Rene Bertrandy, Manager

MISSION: To provide technical assistance to French building companies worldwide.

PRIMARY WORK: Performs work in soil mechanics, laboratory management, civil works and building materials, acoustics and vibrations, and nondestructive testing.

SOURCE OF FINANCES: Government and fees for research.

PUBLICATIONS: Journal, monthly reports, and research reports.

SCIENTIFIC AND TECHNICAL RESEARCH CENTER (CSTB)

4 avenue Recteur Poincare
75782 Paris Cedex 16

CONTACT: Pierre Chemillier, Director

MISSION: National building research laboratory providing the building community with improved building practices.

PRIMARY WORK: Performs research and testing in construction technology, building equipment, HVAC, acoustics, fire safety, economics, architectural research, structural loads, and robotics.

SOURCE OF FINANCES: Government, contract research, and sales of publications.

PUBLICATIONS: Monthly journal, technical standards and codes of practice, and research reports.

WOOD AND FURNITURE TECHNICAL CENTER (CTB)

10 avenue de Saint-Mande
75012 Paris

CONTACT: Daniel Guinard, General Director

MISSION: A primary testing center to encourage the application of wood products and to improve raw material recovery and product quality.

PRIMARY WORK: Provides fundamental and applied studies to improve the knowledge of wood materials; undertakes standardization and quality control activities; manages the French standard (NF) mark and CTB quality labels; provides technical support to industry, links to international organizations, and documentation of information.

SOURCE OF FINANCES: Fees for contracts, sales of publications, and grants.

PUBLICATIONS: Periodic and specialty publications.

TECHNICAL CENTER FOR AIR HANDLING AND HEATING INDUSTRIES

Plateau du Moulon-B.P. 19
91402 Orsay Cedax

CONTACT: Sulejman Becirspahic

MISSION: To promote technical progress and contribute to improving efficiency and quality of air handlers and heating equipment.

PRIMARY WORK: Provides work in heating, combustion, air-handling techniques, energy conservation, calibration, and training.

SOURCE OF FINANCES: Government and fees for services.

PUBLICATIONS: Recommendations, research reports, and specialty publications.

PUBLIC WORKS RESEARCH LABORATORY (LCPC)

58 boulevard Lefebvre
75732 Paris Cedex 15

CONTACT: Jean Frances Coste, Director

MISSION: To serve as central laboratory for roads, structures, bridges, civil engineering materials, and the environment.

PRIMARY WORK: Performs research in the following areas: materials, structures, painting, geotechnical, soil and rock mechanics, water and environment, inspection and pathology of construction works, and documentation of scientific and technical information.

SOURCE OF FINANCES: Government.

PUBLICATIONS: Bimonthly bulletin, recommendations, technical information notes, catalogues, and research reports.

FRENCH CENTRAL HYDRAULICS LABORATORY (LCHF)

rue Eugene Renault 10
94700 Maison-Alfort

CONTACT: P. Prudhomme, General Director

MISSION: To serve private hydraulics engineering and water management organizations.

PRIMARY WORK: Performs work in maritime and river hydraulics and information processing.

SOURCE OF FINANCES: Private grants.

PUBLICATIONS: Research reports and specialized reviews.

QUALITEL

136 Bd Germain
75006 Paris

CONTACT: Claude Trehin, Managing Director

MISSION: To improve the quality of new construction and certify the quality of the technical conception of a project with the "Label Qualitel" trademark. Through a valuation method the trademark considers heat insulation, soundproofing inside and outside the building, installation of electricity, plumbing, frontage, and roof coating.

PRIMARY WORK: Aids professionals to conceive quality projects, value the qualitative content of the projects before definitive construction, allow improvement, deliver Quality Label, and provide consumer information on quality housing.

SOURCE OF FINANCES: Profits.

PUBLICATIONS: *Guide Qualitel* (valuation method) and information seminars.

SYNDICATES OF REINFORCED CONCRETES AND INDUSTRIALIZED TECHNIQUES (SNBATI)

9 rue la Perouse
75784 Paris Cedex 16

CONTACT: Phillipe Levaux, President

MISSION: To stimulate innovation in the civil engineering industry.

PRIMARY WORK: Provides training and research in fire, soils, materials, pathology, concrete, structures, and equipment; documentation and seminars.

SOURCE OF FINANCES: Subscription from member firms.

PUBLICATIONS: Research results, professional journals, syndicate journal, and other specialty publications.

INTERPROFESSIONAL TECHNICAL UNION OF NATIONAL FEDERATIONS OF BUILDING AND PUBLIC WORKS (UTI)

6-14 rue la Perouse
75784 Paris Cedex 16

CONTACT: J. Brunier, President

MISSION: Primary testing laboratory to improve building practices and further the progress of the construction industry.

PRIMARY WORK: Performs research in soils and foundations, acoustics, materials, fire security, and building physics; tests large structural systems and products; operates bibliographic center and documentation center; assists firms to find solutions to technical problems and regulations; and performs testing and training in advanced studies in construction and building physics.

SOURCE OF FINANCES: Fees for services and levy from construction firms.

PUBLICATIONS: Results of research, monthly and bimonthly journals, proceedings, and monographs.

GERMAN DEMOCRATIC REPUBLIC

ACADEMY OF BUILDING OF THE GERMAN DEMOCRATIC REPUBLIC

1123 Berlin
Plauener Strasse

MISSION: To serve as the principal building research institute in West Germany.

PRIMARY WORK: Provides research, development, planning, design, and consulting in architecture, civil engineering, and building construction through its 12 institutes: Town Planning and Architecture; Housing and Related Buildings; Industrial Buildings; Agricultural Buildings; Civil Engineering; Technology and Mechanization; Design and Standardization; Heating, Ventilating, and Structural Theory; Building Materials; Economics; Building Information Center; and Experimental Projects Division.

SOURCE OF FINANCES: Government and fees for services.

PUBLICATIONS: Research reports, technical notes, and specialty publications.

TECHNICAL UNIVERSITY OF DRESDEN, SECTIONS OF CIVIL ENGINEERING, ARCHITECTURE, ENERGY TRANSFORMATION

Mommsenstrasse 13
8027 Dresden

CONTACT: Professor Kurth, Dean

MISSION: To provide building research and advanced education in social, natural, and technical sciences.

PRIMARY WORK: Performs research in civil engineering, architecture, water engineering, computer-aided manufacturing, building diagnostics, heat and moisture transfer, maintenance, modernization, and reconstruction.

SOURCE OF FINANCES: Government, university, and building industry.

PUBLICATIONS: Research reports and professional journals.

GHANA

BUILDING AND ROAD RESEARCH INSTITUTE (BRRI)

University P.O. Box 40
Kumasi

CONTACT: M. D. Gidigasu, Director

MISSION: National building research institute supporting the building and road design and construction industries.

PRIMARY WORK: Performs research in the design and planning of buildings, building and road construction, materials, soil mechanics and foundations, structural systems, concrete and masonry products, transportation and traffic planning; disseminates information.

SOURCE OF FINANCES: Government and fees for services.

PUBLICATIONS: Quarterly newsletter, research reports, brochures, annual reports, and bibliographies.

GREECE

**ARISTOTLE THESSALONIKI UNIVERSITY,
SCHOOL OF ENGINEERING,
DEPARTMENT OF CIVIL ENGINEERING,
DIVISION OF STRUCTURAL ENGINEERING**

Thessaloniki

CONTACT: G. Penelis, Division Director

MISSION: To provide education and civil engineering research.

PRIMARY WORK: Performs research in static and dynamic analysis of structures, reinforced concrete, materials testing, and earthquake engineering.

SOURCE OF FINANCES: University.

PUBLICATIONS: Research reports, conference papers, and professional journals.

**NATIONAL TECHNICAL UNIVERSITY OF ATHENS,
DEPARTMENT OF THEORETICAL AND APPLIED MECHANICS**

5 Heroes of Polytechnion Avenue
Athens (624)

CONTACT: P. S. Theocaris, Director

MISSION: Advanced education and research services to the building community.

PRIMARY WORK: Undertakes research in the mechanics of materials and materials testing, construction use, and quality control.

SOURCE OF FINANCES: University.

PUBLICATIONS: Research results and professional journals.

GUATEMALA

SAN CARLOS UNIVERSITY, ENGINEERING RESEARCH CENTER (CII)

Ciudad Universitaria
Zona 12
Guatemala City

CONTACT: Emilio Beltranena Matheu, Director

MISSION: Education and research services to the building community.

PRIMARY WORK: Performs research in materials, structural engineering, building methods, housing typology, nonconventional energy sources, and sanitary engineering; serves the community with standardization, testing and analysis, and technical assistance services.

SOURCE OF FINANCES: Government, university, fees for contract research, and grants.

PUBLICATIONS: Bimonthly journal, technical standards, research reports, quarterly reference bulletin, monthly lists of publications, and bibliographies.

HUNGARY

CENTRAL RESEARCH AND DESIGN INSTITUTE FOR THE SILICATE INDUSTRY

Becsi ut 126/128
P.O. Box 112
H-1300 Budapest

CONTACT: Jeno Simon, Director

MISSION: Representation of the Hungarian design and engineering organizations for silicate building materials industry.

PRIMARY WORK: Research and development on preparation and production of raw materials, special cements, glasses, ceramics and perlites; performs measurements and designs related to power engineering, energy utilization, air quality, and building materials.

SOURCE OF FINANCES: Fees for research.

PUBLICATIONS: Periodic scientific publications, monographs, and reports.

HUNGARIAN INSTITUTE FOR BUILDING SCIENCE (ETI)

David Ferenc u. 6.
H 1113 Budapest

CONTACT: Sandor Vasas, General Director

MISSION: National building research institute with laboratories providing technology to improve building practice.

PRIMARY WORK: Performs research in structural engineering, building physics, innovative construction methods and techniques, mechanization of building construction and automation, and building diagnostics.

SOURCE OF FINANCES: Government and fees for research.

PUBLICATIONS: Quarterly research report, yearbook, bulletins, scientific proceedings, and proceedings on computer techniques.

INFORMATION CENTER OF BUILDING (ETK)

Harsfa u 21
1074 Budapest, Vll

CONTACT: Peter Hamvay, Director

MISSION: To serve as the central repository of Hungarian building technology documentation.

PRIMARY WORK: Provides technical assistance to the building community in building and documentation services; develops data bases, translations, and exhibitions; provides information about new foreign and domestic products; provides help to private housing efforts; and operates the Eastern-European Office of UN HABITAT.

SOURCE OF FINANCES: Sales of publications and fees for services.

PUBLICATIONS: Weekly bulletins, monthly journals and quarterly reviews of various technical disciplines, catalogues, research reports, and data bases.

INSTITUTE FOR QUALITY CONTROL OF BUILDING (EMI)

Dioszegi ut 37
1113 Budapest Xl

CONTACT: Imre Borbely, Director

MISSION: To serve as government research and testing institute for quality control in the field of the building and building materials industry.

PRIMARY WORK: Performs official quality control and testing of building materials and construction equipment; issues quality control certificates, technical standards and regulations; maintains security control of lifts and building machines; and performs building inspections. The main laboratories focus on fire resistance, thermal physics, structural, mechanical, and chemical.

SOURCE OF FINANCES: Government and fees for services.

PUBLICATIONS: Bimonthly journal on building quality, research reports, list of publications, and standards and guidelines.

TECHNICAL UNIVERSITY OF BUDAPEST, FACULTY OF CIVIL ENGINEERING, FACULTY OF ARCHITECTURE

Muegyetem rkp 3 1 em
1111 Budapest

CONTACT: Otto Halasz, Dean of Civil Engineering
Gyorgy Deak, Dean of Architecture

MISSION: To serve as institution of higher education; the only institution that graduates certified civil engineers and architects.

PRIMARY WORK: Performs research on construction and architecture, structural engineering and steel structures, geotechnics, road engineering, reinforced concrete, building materials, building operations and equipment, sanitary engineering, housing, and town planning.

SOURCE OF FINANCES: Government and university.

PUBLICATIONS: Research reports and journal articles.

ICELAND

BUILDING RESEARCH INSTITUTE

Keldnaholt
110 Reykjavik

CONTACT: Hakon Olafsson, Director

MISSION: To serve as the government research and testing laboratory.

PRIMARY WORK: Performs work in the areas of concrete, structural engineering, building technology, soil mechanics, road construction; performs cost analysis; and undertakes information documentation.

SOURCE OF FINANCES: Government.

PUBLICATIONS: Technical recommendations and research reports.

INDIA

CENTRAL BUILDING RESEARCH INSTITUTE (CBRI)

Roorkee (UP)

CONTACT: R. K. Bhandari, Director

MISSION: National building research laboratory that provides the building industry with knowledge to solve building design and construction problems.

PRIMARY WORK: Performs research on building materials, geotechnical engineering, architectural and physical planning, building plants, processes and productivity, construction, fire, fire characterization and safety, rural buildings; and develops and implements new techniques for building efficiencies.

SOURCE OF FINANCES: Government.

PUBLICATIONS: Building digests, data sheets, abstracts, annual reports, proceedings of conferences, special publications, and newsletter.

CEMENT RESEARCH INSTITUTE OF INDIA (CRI)

Delhi-Mathura Road
PO CRRI
New Delhi 110 020

CONTACT: M. P. Dhir, Director

MISSION: National laboratory performing basic and applied research to solve national road problems.

PRIMARY WORK: Performs research on construction materials, transportation engineering and road characteristics, highway infrastructure; and provides technical assistance for planning, training, and documenting technology.

SOURCES OF FINANCE: Government.

PUBLICATIONS: Research reports, technical papers, annual reports, abstracts, and catalogues.

INDIAN CONCRETE INSTITUTE

CSIR Campus
Madras-600 113

CONTACT: Zacharia George, Secretary-General

MISSION: To advance the state of the art of concrete and concrete products for the construction industry.

PRIMARY WORK: Performs work in concrete and ferrocement; conducts training seminars; and advises industry.

SOURCE OF FINANCES: Government, fees for services, and membership subscription.

PUBLICATIONS: Journal, annual report, and special publications.

INDIAN INSTITUTE OF TECHNOLOGY, KANPUR (IIT, KANPUR)

IIT post Office
Kanpur 208 016

CONTACT: S. Sampath, Director

MISSION: Education in engineering and research expertise.

PRIMARY WORK: Undertakes research in civil, environmental, hydraulics, geotechnical, geological, and transportation engineering; and materials sciences, electronics and laser technology.

SOURCE OF FINANCES: University and government.

PUBLICATIONS: Research reports, annual report, and specialty reports.

NATIONAL BUILDINGS ORGANIZATION (NBO)

G Wing
Nirman Bhawan
Maulana Azad Road
New Delhi

CONTACT: G. C. Mathur, Director

MISSION: To promote experimental housing projects, stimulate improved quality control, and serve as a research and development institution.

PRIMARY WORK: Performs research in design and construction techniques for low-cost housing, provides documentation and technology transfer expertise, and serves as UN Housing Regional Housing Center.

SOURCE OF FINANCES: Government.

PUBLICATIONS: Semiannual journal, research reports, handbooks, directories, and other specialty publications.

NATIONAL INSTITUTE OF CONSTRUCTION MANAGEMENT AND RESEARCH (NICMAR)

Walchand Center
Walchand Terraces
Tardeo Road
Bombay 400 034

CONTACT: G. R. Dole, Deputy Registrar

MISSION: To promote and upgrade the managerial capabilities and technical competency of the Indian construction industry.

PRIMARY WORK: Conducts multidisciplinary education, training, research, and consultancy programs.

SOURCE OF FINANCES: Member contributions, annual subscriptions, tuition and consultancy fees, and contribution to research projects.

PUBLICATIONS: Quarterly journal, occasional publications, and bibliographies.

STRUCTURAL ENGINEERING RESEARCH CENTER, MADRAS (SERC, MADRAS)

CSIR Campus
Taramani, Madras 600 113

CONTACT: M. Ramsiah, Director

MISSION: To provide research in structural engineering, housing, and technical assistance to the building community.

PRIMARY WORK: Performs research in computer-aided design, off-shore structures, structural engineering, materials sciences, reinforced concrete, concrete composites, concrete construction techniques, and disaster-resistant buildings.

SOURCE OF FINANCES: Government.

PUBLICATIONS: Quarterly journal of structural engineering, designs for low-cost housing, publications in professional journals, annual report, and specialty publications.

INDONESIA

DIRECTORATE OF BUILDING RESEARCH (DBR), CENTER FOR RESEARCH AND DEVELOPMENT ON HUMAN SETTLEMENTS (DPMB)

84 Jalan Tamansari (Tromol Pos 15)
Bandung

CONTACT: Karman Somawidjaja, Director

MISSION: To serve as a regional center for research on housing concepts, building materials, structures, and human settlements.

PRIMARY WORK: Performs research on evaluation of building materials, structures, and construction techniques; planning and development of human settlements; training; information processing; and documentation.

SOURCE OF FINANCES: Government and foreign grants.

PUBLICATIONS: Quarterly journal, monthly abstracts, technical standards, research reports, and specialty publications.

INDONESIAN ROAD RESEARCH INSTITUTE (IRRI)

Jalan Raya Timur No. 264
P.O. Box 298
Bandung

CONTACT: H. A. B. Hasibuan, Director

MISSION: To serve the building industry with central resources for road research and improved construction and maintenance techniques.

PRIMARY WORK: Performs research in structural and materials engineering, concrete technologies, testing and measurement techniques, and classification and documentation methods.

SOURCE OF FINANCES: Government.

PUBLICATIONS: Monthly journal, technical standards, and research reports.

IRAQ

NATIONAL CENTER FOR CONSTRUCTION LABORATORIES (NCCL)

Mousa Bin Naseer Square
Tell Mohammed
Baghdad

CONTACT: Mufid A. Samarai, Director General

MISSION: National building research laboratory that serves the building community through advancing building technology.

PRIMARY WORK: Performs research in construction materials, concrete, stone, foundations, subsoils; conducts laboratory tests for quality control; and undertakes documentation activities.

SOURCE OF FINANCES: Government.

PUBLICATIONS: Standards, quality control of construction materials, and specialty reports.

IRELAND

INSTITUTE FOR INDUSTRIAL RESEARCH AND STANDARDS, CONSTRUCTION INDUSTRY DIVISION (IIRS)

Ballymun Road
Dublin 9

CONTACT: S. F. Dunleavy, Director, Construction Industry Division

MISSION: Responsible for setting standards, operating technical certification programs, providing test facilities, and performing building research that improves construction industry technologies and quality.

PRIMARY WORK: Performs research on materials, civil and structural engineering, foundations, concrete, construction techniques and technologies, energy conservation, environmental engineering, product and process development, quality assurance, and certification.

SOURCE OF FINANCES: Government, fees for services, and grants.

PUBLICATIONS: Bimonthly journal, standards, publishes in technical journals, proceedings of conferences, and specialty publications.

NATIONAL INSTITUTE FOR PHYSICAL PLANNING AND CONSTRUCTION RESEARCH (AFF)

St. Martin's House
Waterloo Road
Dublin 4

CONTACT: L. M. McCumiskey, Chief Executive Officer

MISSION: To provide technical support to advance building technology for implementation of national policies in the area of planning, social, and economic aspects.

PRIMARY WORK: Performs research in environmental engineering, housing, specifications, water supply and disposal, physical planning, preservation, development of national physical infrastructure, road and building construction; and training and information activities.

SOURCE OF FINANCES: Government and fees for services.

PUBLICATIONS: Semiannual journal, research reports, and specialty publications.

UNIVERSITY OF DUBLIN, TRINITY COLLEGE

Dublin 2

CONTACT: W. A. Watts, The Provost

MISSION: Education and research to advance building practices.

PRIMARY WORK: Performs research in structural engineering, soil mechanics, thermal engineering, and materials.

SOURCE OF FINANCES: Government.

PUBLICATIONS: Results published in professional journals.

ISRAEL

BUILDING RESEARCH STATION (BRS)

Technion-Israel Institute Of Technology
Technion City
Haifa 32 000

CONTACT: A. Warszawski, Head

MISSION: National building research laboratory that develops technology to advance building practices.

PRIMARY WORK: Performs research in building materials and technology, concrete, civil and earthquake engineering, construction management, economics, robotics, expert systems, CAD in buildings, climatology, acoustics, quality control and assurance; provides testing, cost control, information management and dissemination; and oversees civil engineering dissertations.

SOURCE OF FINANCES: Government, cement foundation, and building industry.

PUBLICATIONS: Monthly journal, standards, research reports, and bibliographies.

TRANSPORTATION RESEARCH INSTITUTE (TRI)

Technion City
Haifa 32 000

CONTACT: Daniel Sheder, Director

MISSION: Serve as central source for transportation research.

PRIMARY WORK: Performs research in road and traffic engineering, energy use and conservation, planning and land use, and road safety.

SOURCE OF FINANCES: Government.

PUBLICATIONS: Research reports, quarterly bulletin, biannual activities report, and annual research program.

ITALY

RESEARCH INSTITUTE FOR LIGHT METALS (ISML)

Via G. Fauser 4
28100 Novara

CONTACT: Paola Fiorine, Director

MISSION: To serve as the center for aluminum research and technology.

PRIMARY WORK: Performs research in metals, surface treatments, engineering of new products, and characteristics and performance of light alloys.

SOURCE OF FINANCES: Government and research contracts.

PUBLICATIONS: Monographs, research published in professional journals, and annual directory of publications.

ITALIAN TECHNICAL AND ECONOMIC CEMENT ASSOCIATION

Via di S. Theresa 23
00198 Rome

CONTACT: Mario Manicardi, Director

MISSION: To promote the application of cement by providing technology and testing services to association members.

PRIMARY WORK: Performs work in concrete technology, economics, standards development, information development, and dissemination.

SOURCE OF FINANCES: Members fees and annual tax on cement production.

PUBLICATIONS: Monthly magazine, quarterly journal, brochures, documentation services, and bibliography.

RESEARCH AND EXPERIMENTATION CENTER FOR THE CERAMIC INDUSTRY

Via Martelli 26
40138 Bologna

CONTACT: Carlo Palmonari, Director

MISSION: To promote improved ceramic products through technology to the ceramic industry.

PRIMARY WORK: Performs research on ceramic, tile, masonry materials; undertakes tests and analysis of raw materials; provides education and training; and holds meetings and seminars.

SOURCE OF FINANCES: Industry and member organizations.

PUBLICATIONS: Bimonthly publication and research reports.

BUILDING CENTER

Via Rivoltana 8
20090 Segrate
Milano

CONTACT: Umberto Sportelli, Managing Director

MISSION: Advanced services for the construction industry.

PRIMARY WORK: Building sector market analysis, technological innovation, technical information on building products, courses, seminars, professional training, and integrated building intervention projects.

SOURCE OF FINANCES: Private.

PUBLICATIONS: Standards and proceedings of meetings.

NATIONAL ASSOCIATION OF ITALIAN ENGINEERS AND ARCHITECTS (ANIAI)

Consiglio Federale
Piazza Sallustio 24
00187 Rome

MISSION: To offer member architects and engineers improved building practices.

PRIMARY WORK: Provides resource for regulations on design and construction, environmental protection, and training.

SOURCE OF FINANCES: Member fees.

PUBLICATIONS: Regulations, codes of practice, and bulletins.

NATIONAL RESEARCH COUNCIL INSTITUTE FOR HOUSING AND SOCIAL INFRASTRUCTURES (IRIS)

Via Crocifisso 2/b
70126 Mungivacca
Bari

CONTACT: Giovanni Tortorici, Director

MISSION: National building research institute.

PRIMARY WORK: Performs research on building materials and components; building failures; renovation, maintenance, and preservation; town planning; economics; and advanced testing techniques.

SOURCE OF FINANCES: Government.

PUBLICATIONS: Annual report and articles published in professional journals.

CENTRAL INSTITUTE FOR INDUSTRIALIZATION AND BUILDING TECHNOLOGY (ICITE)

Via Lombardia 49
Frazione Sesto Ulteriano
20098 San Giuliano Milanese

CONTACT: M. Ferri, Director

MISSION: Standardization and control of building components and systems.

PRIMARY WORK: Performs work in standardization and industrialization of building, Agrement, acoustics, ventilation, thermal insulation, concrete, social housing, energy conservation, building data bases, testing, and quality marks.

SOURCE OF FINANCES: Government.

PUBLICATIONS: Monographs and specialty publications.

BUILDING AND CIVIL ENGINEERING TESTS AND RESEARCH INSTITUTE (ISTEDIL)

Via Tiburtina Valeria km 18 300
00012 Guidonia Montecelio
P.O. Box 7237
00100 Roma Momentano

CONTACT: Manfredo Manfredi, General Manager

MISSION: To perform building control tests and certification.

PRIMARY WORK: Performs work in civil engineering and building technology, provides testing services, agreements for materials and components, and conducts courses and seminars.

SOURCE OF FINANCES: Fees for services.

PUBLICATIONS: Research reports and bibliographies.

FLORENCE UNIVERSITY, FACULTY OF ARCHITECTURE, CONSTRUCTION DEPARTMENT, DEPARTMENT OF PROCESSES AND METHODS IN BUILDING PRODUCTION

Piazza Brunelleschi 6
50121 Firenze

CONTACT: Salvatore Di Pasquale, Director of Construction Department
Arch Romano Del Nord, Director of Building Production Department

MISSION: To provide education and research to advance building practices. The Construction Department, together with the Faculties of Architecture of Rome and Genoa arranged a postgraduate course to give a Ph.D. degree in Structural Analysis in Architecture.

PRIMARY WORK: Performs research in structural engineering, computational mechanics, masonry structural analysis, bridge and large structures, earthquake engineering, geotechnics and foundation engineering, methods for the control of building processes, fabrication and construction techniques, industrialized products.

SOURCE OF FINANCES: University, government, and fees for services.

PUBLICATIONS: Research reports, monographs, and specialty publications.

UNIVERSITY OF NAPLES, DEPARTMENT OF GRAPHIC REPRESENTATION AND DESIGN IMPLEMENTATION

Via Monteoliveto 3
80134 Napels

CONTACT: Virginai Gangemi, Head

MISSION: To educate and conduct research to advance building practices.

PRIMARY WORK: Performs research in architectural and industrial design, graphic representation, control of materials, and building procedures.

SOURCE OF FINANCES: University and fees for research.

PUBLICATIONS: Research reports, proceedings, and specialty publications.

ROME UNIVERSITY,
DEPARTMENT OF INDUSTRIAL DESIGN AND BUILDING PRODUCTION

53 Via Antonio Gramsci
00197 Rome

CONTACT: Eduardo Vittoria, Director

MISSION: Education and research to advance technological innovation and design for the building design, industrial design, and environmental design.

PRIMARY WORK: Performs research on rehabilitation and restoration, quality assurance and control, utility furniture, design of urban open spaces, conceptual project, standards and regulations.

SOURCE OF FINANCES: University and fees for research.

PUBLICATIONS: Research results, professional journals, proceedings, and other specialty publications.

BARI UNIVERSITY, INSTITUTE OF ENGINES AND ENERGETICS, LABORATORY OF THERMOENERGETICS

Via Re David 200
I 70125 Bari

CONTACT: Paolo Bondi

MISSION: To serve as an educational institution and research laboratory to advance building practices.

PRIMARY WORK: Performs research on thermal properties of materials (insulation, vapor transmission, durability, capacity) and energy conservation; also serves as a connection with national and international standards upgrading activities.

SOURCE OF FINANCES: Government and industry

PUBLICATIONS: Research reports and specialty publications.

**BARI UNIVERSITY,
INSTITUTE OF TECHNICAL PHYSICS AND THERMOTECHNICAL PLANTS**

Via re David 200
I 70125 Bari

CONTACT: Renato Lazzarin, Director

MISSION: To serve as an educational institute and research laboratory to advance building practices.

PRIMARY WORK: Performs research in thermal engineering and insulation, energy conservation, acoustics, and meteorology.

SOURCE OF FINANCES: Government and industry.

PUBLICATIONS: Research reports and specialty publications.

JAMAICA

JAMAICA BUILDING RESEARCH INSTITUTE

Medi Center Building
34 Old Hope Road
P.O. Box 505
Kingston 5

CONTACT: Keith Gilfillian, Director

MISSION: National research laboratory to advance building practices.

PRIMARY WORK: Performs research on soils and clay engineering; materials including stones, bricks, and tiles; low-cost roofing; indigenous and waste materials; and infrastructure development.

SOURCE OF FINANCES: Government and United Nations.

PUBLICATIONS: Research reports, annual publication, and specialty publications.

JAPAN

BUILDING CENTER OF JAPAN (BCJ)

Nihon Kenchiku Senta
Mori Building 30
2-2 Toranonion 3-choine
Minato-ku
Tokyo 105

CONTACT: Mitsufusa Sawada, President
                    Yujiro Kaneko, Senior Executive Director

MISSION: To provide building technology to building associations and contractors.

PRIMARY WORK: Performs work on developing standards and specifications, information systems and dissemination methods, operates exhibition center, and performs technical evaluations.

SOURCE OF FINANCES: Fees for information services, technical certification, sales of publications, and contract research.

PUBLICATIONS: Monthly journal, research reports, committee reports, and Building Standard Law.

BUILDING CONTRACTORS SOCIETY (BCS)

Kenchikugyo Kyokai
2-5-1 Hatchobori
Chuo-Ku
Tokyo 104

CONTACT: Hajime Sako, Chairman

MISSION: To serve as a center for building research for member companies.

PRIMARY WORK: Awards the BCS Prize; performs investigations for safety and health-related issues; conducts research on machinery, energy, materials and construction techniques, information systems and dissemination, and scholarships.

SOURCE OF FINANCES: Member companies.

PUBLICATIONS: Quarterly journal, research reports, BCS's annual building prize, and specialty publications.

BUILDING RESEARCH INSTITUTE (BRI)

Kensetsu-Sho Kenchiku Kenkyujo
1 Tatehara, Oho-Machi
Tsukuba-Gun, Ibaraki-Prefecture

CONTACT: Hiroshi Takebayashi, Director

MISSION: National building research laboratory that advances building practices through improved research.

PRIMARY WORK: Performs research in disaster mitigation, structural engineering, materials, energy conservation and energy utilization, housing and building economy, fire engineering, environmental design, and testing and evaluation.

SOURCE OF FINANCES: Government.

PUBLICATIONS: Research reports, bibliographies, and specialty publications.

HAZAMA-GUMI TECHNICAL RESEARCH INSTITUTE

Kabushiki Kaisha Hazama-Gumi Gijutsu Kenkyujo
17-23 Hon-Machi Nishi 4-Chrome
Yono, Saitama 338

CONTACT: Keiichi Fujita, Director

MISSION: To conduct research to improve construction technology for Hazama-Gumi Construction Corporation.

PRIMARY WORK: Performs research in civil engineering, structural and seismic analysis, foundations and soils mechanics, ocean engineering, construction materials, construction equipment, energy-related research, and construction management.

SOURCE OF FINANCES: Hazama-Gumi Construction Corporation.

PUBLICATIONS: Annual journal, research reports, bibliographies, and catalogues.

KAJIMA INSTITUTE OF CONSTRUCTION TECHNOLOGY (KICT)

Kajima Kensetsu Gijutsu Kenkyusho
19- 1, Tobitakyu 2-chome
Chofu-shi, Tokyo 182

CONTACT: Sukenobu Momoshima, Director

MISSION: Technical research laboratory of Kajima Corporation.

PRIMARY WORK: Performs research in structural engineering; soils; foundations; rock mechanics; materials; construction methods, control, and equipment; environmental engineering; piping and air conditioning; and hydraulics.

SOURCE OF FINANCES: Kajima Corporation.

PUBLICATIONS: Annual report and research reports.

MAEDA CONSTRUCTION TECHNICAL RESEARCH INSTITUTE

10-26, Fujimi 2-chrome Chiyoda-ku
Tokyo

CONTACT: Hiroyuki Kitamura, Director

MISSION: To perform research to solve public works construction problems and social needs, and to improve the quality of construction.

PRIMARY WORK: Provides research in civil, soil and geotechnical engineering, materials, construction methods, and environmental engineering.

SOURCE OF FINANCES: Maeda Construction Company.

PUBLICATIONS: Annual journal, research reports, bibliographies, and specialty publications.

NIPPON TELEGRAPH AND TELEPHONE CORPORATION, BUILDING ENGINEERING DEPARTMENT

Nippon Denshin Denwa Kabushiki-gaisya Kenchiku-Bu
1-5-3, Ohte-machi
Chiyoda-ku
Tokyo 100

CONTACT: Kiyoshi Morishita, Executive Manager

MISSION: To provide the telecommunications services with research to improve building technology.

PRIMARY WORK: Construction, maintenance, and restoration and management of telecommunications buildings and facilities.

SOURCE OF FINANCES: Private company.

PUBLICATIONS: Bimonthly journal, intelligent building design manual, and telecommunications buildings standards.

OHBAYASHI-GUMI, LTD.

3, 2-chrome
Kanda Tsukasa-cho, Chiyoda-ku
Tokyo 101

MISSION: To provide technology that supports Ohbayashi-Gumi's construction activities.

PRIMARY WORK: Performs research in civil engineering, soils and foundations, earthquakes, structural materials, construction and environmental technology, and acoustics.

SOURCE OF FINANCES: Ohbayashiu-Gumi Ltd.

PUBLICATIONS: Biannual research reports.

SATO KOGYO CO, LTD.,
CENTRAL TECHNICAL RESEARCH INSTITUTE

Chuo Gijutsu Kenkyusho
47-3 Sanda, Atsugi-shi
Kanagawa Prefecture

CONTACT: Kakuichi, General Manager

MISSION: To provide improved construction technologies that meet the needs of the building industry.

PRIMARY WORK: Research in quality control of concrete, earthquake engineering, subsoil construction engineering, energy conservation, pollution and industrial waste, and testing.

SOURCE OF FINANCES: Fees for services.

PUBLICATIONS: Annual journal.

SHIMIZU CONSTRUCTION CO., LTD.,
INSTITUTE OF TECHNOLOGY

Shimizu-Kensetsu Gijyutsu Kenkyujo
4-17, Etchujima 3-chrome
Koto-ku, Tokyo 135

CONTACT: Toshihiko Ota, Director, General Manager

MISSION: To serve the corporation by developing new construction technologies and improving quality control.

PRIMARY WORK: Performs research in construction methods, architectural performance, material properties, concrete, finishing materials, construction management, steel structures, earthquake engineering, civil and structural engineering, rock mechanics, subsurface development, soil dynamics, seismic engineering, foundation engineering, soil engineering, acoustics, thermal environment, air technology, wind engineering, ocean hydrodynamics, water environment, fire safety, information systems, architectural design methods, electronics, advanced materials, applied biotechnology, and radiation physics.

SOURCE OF FINANCES: Shimizu Construction Company

PUBLICATIONS: Biannual reports of research, *Shimizu Technical Research Bulletin* (annually), and special publications.

SOCIETY OF HEATING, AIR-CONDITIONING, AND SANITARY ENGINEERS OF JAPAN (SHASE)

Kuki Chowa Eisei Kogakkai
Nakajima Building, 8-1
1-chrome Kita-shinjuku, Shinjuku-ku
Tokyo

CONTACT: Shoichi Fujii, President

MISSION: To provide improved building services and environmental engineering technology to the building community.

PRIMARY WORK: Performs research in HVAC and plumbing systems, hosts technical meetings, conducts training, and presents annual SHASE award for technical excellence.

SOURCE OF FINANCES: Membership fees and research contracts.

PUBLICATIONS: Monthly journal, handbooks, standards, research reports, bibliographies, and other specialty publications.

TAISEI CORPORATION, ENGINEERING AND CONSTRUCTION

25-1, Nishi-Shinjuku 1-chrome
Shinguku-ku
Tokyo

CONTACT: Hajime Sako, President

MISSION: To promote development of advanced design and construction technologies.

PRIMARY WORK: Performs work in construction engineering; building materials; structural engineering including earthquakes, foundations, geotechnical, and hydraulics; building computer control systems; computer-aided construction; environmental and acoustical engineering; pollution control; and collection, management, and dissemination of information.

SOURCE OF FINANCES: Taisei Corporation.

PUBLICATIONS: Research reports and annual report.

TAKENAKA KOMUTEN COMPANY, LTD., TECHNICAL RESEARCH LABORATORY

5-14, 2-chrome
Minamisuna, Koto-ku
Tokyo

CONTACT: Motoki Kondo, Director, General Manager

MISSION: To serve the Takenaka Komuten Co. and the Takenaka Doboku Co. with advanced building and civil engineering technologies.

PRIMARY WORK: Conduct research in systems engineering; disaster prevention and fire engineering, acoustics and vibration engineering; general structural and wind engineering; advanced structures and structural dynamics; computational engineering; ground improvement and soil dynamics; foundation, marine, civil, environmental, and mechanical engineering; water control engineering; concrete engineering; building materials; and construction equipment and materials.

SOURCE OF FINANCES: Takenaka Komuten Co., Takenaka Doboku Co., and government organizations.

PUBLICATIONS: Biannual research reports.

TODA CONSTRUCTION CO., LTD., INSTITUTE OF TECHNOLOGY

Toda Kensetsu Gijutsu Kenkyusho
7-1 Kyobashi 1-chrome
Chuo-ku, Tokyo 104

CONTACT: K. Watanabe, General Manager

MISSION: To serve the Toda Construction Co. and clients with advanced technology for improved construction practices.

PRIMARY WORK: Conducts research in soil, foundation, structural, material, and construction engineering; bedrock analysis; construction methods; and environmental design.

SOURCE OF FINANCES: Toda Construction Co., Ltd.

PUBLICATIONS: Annual research report, quarterly reports, standards, research reports, and bibliographies.

TOHOKU UNIVERSITY, FACULTY OF ENGINEERING, DEPARTMENT OF ARCHITECTURE

Aramaki-Aoba
Sendai 980

CONTACT: K. Hirai, Director

MISSION: To undertake research in architecture and building engineering.

PRIMARY WORK: Performs research in architectural and urban planning, environmental and structural mechanics and engineering, building materials, and architectural history and design.

SOURCE OF FINANCES: Government.

PUBLICATIONS: Annual research reports and other specialty publications.

TOKYO CONSTRUCTION CO., LTD., TECHNICAL RESEARCH CENTER

Tokyu Kensetsu Gijutsu Kenkysho
13-9, Miyazaki 2-chrome
Miyamae-ku
Kawasaki City 213

CONTACT: Kunio Nakamura, General Manager

MISSION: To serve the Tokyu Construction Co. with improved technology to advance construction practices and the state of knowledge for designers.

PRIMARY WORK: Provides research in civil engineering structures, architectural design, construction technologies, and environmental engineering.

SOURCE OF FINANCES: Tokyu Construction Co., Ltd.

PUBLICATIONS: Annual technical report and specialty publications.

JORDAN

BUILDING RESEARCH CENTRE

Royal Scientific Society
P.O. BOX 925819
Amman

CONTACT: Daoud Jabaji, Director

MISSION: To conduct applied research and studies and to provide technical and laboratory services in the fields of building, construction, and road technologies and materials; to support the development and growth of the building industry in Jordan.

PRIMARY WORK: Conducts research and development to develop low-cost building technology; drafts standards, specifications, and codes; quality control and testing of building materials and products, and asphalt mixes; site investigation services and CAD services; and publication of manuals on design and construction.

SOURCE OF FINANCES: Government and private.

PUBLICATIONS: Technical reports, research reports, data on building materials and products, standards, and codes and manuals.

UNIVERSITY OF JORDAN, FACULTY OF ENGINEERING AND TECHNOLOGY, DEPARTMENT OF CIVIL ENGINEERING

University of Jordan
Faculty of Engineering and Technology
Department of Civil Engineering
Amman

MISSION: To promote the use of building materials and the study of related new technology.

PRIMARY WORK: Performs research and consultation as well as conferences and short courses on building technology.

SOURCE OF FINANCES: University of Jordan research funds.

PUBLICATIONS: Quarterly journal.

KENYA

HOUSING RESEARCH AND DEVELOPMENT UNIT

University of Nairobi
P.O. Box 30197
Nairobi

MISSION: To serve the building community with technology to improve building practices.

PRIMARY WORK: Performs research in low-cost housing, building systems, construction materials and techniques; information management and dissemination.

SOURCE OF FINANCES: Government, university, and sale of publications.

PUBLICATIONS: Research reports, construction guidelines, and bibliography.

KOREA

ARCHITECTURAL INSTITUTE OF KOREA (AIK)

No. 2-7, 2-Ka
Myung-Dong, Chung-Ku
Seoul 100

CONTACT: Keun-Duck Kim, President

MISSION: To serve as the central source for information on the building process and building standards.

PRIMARY WORK: Performs work in architectural design and planning, structural engineering, building materials, construction technology, prefabricated buildings, energy conservation, and information systems.

SOURCE OF FINANCES: Fees for services, membership fees, and grants and contributions.

PUBLICATIONS: Bimonthly journal, standards and specifications, research reports, bibliographies, and specialty publications.

KOREA INSTITUTE OF CONSTRUCTION TECHNOLOGY (KICT)

(Life B/D), 61, Yoido-Dong
Yongdungpo-Gu
Seoul 150

CONTACT: Yoon Yong Nam, President

MISSION: To carry out research and development of construction technologies in a systematic and multidisciplinary manner and to make contributions to structural, construction, geotechnical, water resources, environmental, architecture and facilities engineering fields by introducing advanced technologies and promoting widespread dissemination of such technologies to the Korean engineering and construction industry.

PRIMARY WORK: Research and dissemination of advanced construction technologies, new construction methods, construction design and management, construction equipment and materials, construction standards; technical service and consulting activities on construction research; and other research activities to promote the construction industry.

SOURCE OF FINANCES: Ministry of Construction and Contractors Financial Cooperative.

PUBLICATIONS: Monthly journal, annual report, bibliographies, and research reports.

NETHERLANDS

BOUWCENTRUM

Weena 700
P.O. Box 299
3000 AG, Rotterdam

CONTACT: Heinrich Meijer, Director

MISSION: To develop technology to improve, maintain, and renovate the built environment.

PRIMARY WORK: Provides technical, physical, and management research and consultations in dwellings; information systems; training; and exhibitions.

SOURCE OF FINANCES: Research grants, fees for services, sale of publications, and exhibitions.

PUBLICATIONS: Monthly and bimonthly journals, research reports, bibliographies, and specialty publications.

BUILDING RESEARCH FOUNDATION

Weena 740
P.O. Box 20740
30001 JA Rotterdam

CONTACT: J. J. de Bruijn, General Director

MISSION: To represent the building industry with information for improving building practices.

PRIMARY WORK: Provides work in building techniques including structures, materials, construction methods, building failures, city planning, environmental issues, business management problems on-site, organization, economics; transfer of knowledge to the building industry and the building educational and training centers.

SOURCE OF FINANCES: Contractors and building industry research fund.

PUBLICATIONS: Research reports.

TIMBER CENTER FOR RESEARCH, INFORMATION, AND EDUCATION

P.O. Box 401
1400 AK Bussum

CONTACT: A. van der Velden, Managing Director

MISSION: To initiate, stimulate, and coordinate technology development and delivery for improved wood building practices.

PRIMARY WORK: Provides research in material properties, timber structural failures, and information systems.

SOURCE OF FINANCES: Timber trade and industry, sale of publications, and fees for contracts.

PUBLICATIONS: Research reports and list of publications.

CIAD ASSOCIATION FOR COMPUTER APPLICATION IN ENGINEERING

P.O. Box 74
2700 AB Zoetermeer

CONTACT: J. C. le Clercq, Director

MISSION: To represent members through a central source of computer integrated architectural and design information.

PRIMARY WORK: Provides information collection and delivery services and conducts training.

SOURCE OF FINANCES: Membership fees and government subsidies.

PUBLICATIONS: Monographs and specialty publications.

DELFT HYDRAULICS LABORATORY (DHL)

Waterloopkundig Laboratory
Rotterdamseweg 185
P.O. Box 117
2600 MH Delft

CONTACT: J. E. Prins, Director

MISSION: To serve as consulting and research institute in hydrology engineering.

PRIMARY WORK: Performs research in model investigations, design of hydraulic structures, river and coastal engineering; and consulting and advisory services.

SOURCE OF FINANCES: Fees from contracts.

PUBLICATIONS: Quarterly journal, research reports, and specialty publications.

DELFT GEOTECHNICS

Stieltjesweg 2
P.O. Box 69
2600 AB Delft

CONTACT: W. Stevelink, Managing Director

MISSION: To serve as a consulting and research institute in the field of soil mechanics, foundation engineering, and environmental geotechnology.

PRIMARY WORK: Performs theoretical and laboratory research in geotechnical engineering and environmental geotechnics; offers testing services in the field, in the laboratory, and in a geotechnical centrifuge; and also offers information systems.

SOURCE OF FINANCES: Fees for research and consultancy.

PUBLICATIONS: Semiannual journals and bimonthly abstracts.

DELFT UNIVERSITY OF TECHNOLOGY, DEPARTMENT OF CIVIL ENGINEERING

P.O. Box 5048
2600 GA Delft

CONTACT: G. J. van Alphen, Manager, Department of Civil Engineering

MISSION: To undertake research in civil engineering and town planning for improving building practices.

PRIMARY WORK: Performs research in civil engineering, materials, environmental engineering, road and railroad construction, fluid mechanics.

SOURCE OF FINANCES: University.

PUBLICATIONS: Research reports, semiannual information reports, and specialty publications.

DUTCH CEMENT INDUSTRY ASSOCIATION

P.O. Box 3011
5203 DA, 's-Hertogenbosch

CONTACT: C. F. A. Wolterbeek, Director

MISSION: To stimulate a wider and better use of cement and concrete.

PRIMARY WORK: Undertakes market development activities, promotes use of concrete, and collects and disseminates information on application of concrete.

SOURCE OF FINANCES: Dutch cement manufactures.

PUBLICATIONS: Monthly, bimonthly, and quarterly journals; books on concrete constructions and applications; market development leaflets and brochures; and list of publications.

**EINDHOVEN UNIVERSITY OF TECHNOLOGY,
DEPARTMENT OF ARCHITECTURE, BUILDING, AND PLANNING,
MATERIALS AND STRUCTURES LABORATORY**

P.O. Box 513
5600 MB Eindhoven

CONTACT: E. M. M. G. Niel
J. W. B. Stark

MISSION: To provide education and research to advance building practices.

PRIMARY WORK: Performs research in materials including masonry, timber, metals, and waste materials; civil engineering, including off-shore structures, fracture mechanics, development of new construction, innovations, and building re-use.

SOURCE OF FINANCES: Government.

PUBLICATIONS: Standards, research reports, and specialty journals.

FOUNDATION FOR ARCHITECTS' RESEARCH

Eindhoven University,
P.O. Box 429
Eindhoven

CONTACT: John C. Carp

MISSION: To advance knowledge of the decision making, design, and production process for improved usability of the built environment.

PRIMARY WORK: Performs research in the development of design and decision methods for housing, spatial relationships, and user control; promotes real-world applications and exchange of professional experience through worldwide SAR-NETWORK.

SOURCE OF FINANCES: Donations by sponsors and fees for research and services.

PUBLICATIONS: Specialty publications.

**QUALITY DECLARATIONS ORGANIZATIONS FOR BUILDING MATERIALS
AND COMPONENTS (KOMO)**

Sir Winston Churchill-Laan 273
P.O. Box 240
2280 AE Rijswijk ZH

CONTACT: F. Wagenmaker, Director

MISSION: To serve as the central institute for quality declarations (certificates, conformity-marks and agrements) for the building industry.

PRIMARY WORK: Performs tests, assesses the results and issues certificates and agrements for building materials and structural components, quality control, regulations, testing, and information.

SOURCE OF FINANCES: Fees for services and sales of publications.

PUBLICATIONS: Quarterly information bulletins, annual report, and technical criteria.

L.M.I. B.V. LABORATORIES FOR MATERIALS AND INDUSTRIES

Kerkrade
P.O. Box 134
Rimburgerweg 2
6471 XX Eygelshoven

CONTACT: J. Hoogman
          W. van Alem

MISSION: Research and development for industries.

PRIMARY WORK: Performs research on waste products, recycling of building materials, new utilizations for coal ash and gypsum, natural and artificial stones, metals corrosion, and environmental impact on building materials.

SOURCE OF FINANCES: Industry, government, and private investment.

PUBLICATIONS: Research reports and laboratory results.

NATIONAL HOUSING ASSOCIATION

Markenlaan 1
P.O. Box 50051
1305 AB Almere

CONTACT: B. G. A. Kempen, General Director

MISSION: To serve housing associations with knowledge about improved operation and maintenance of housing.

PRIMARY WORK: Provides services in management and organization, technical developments, and financial aspects.

SOURCE OF FINANCES: Fees from members and services.

PUBLICATIONS: Biweekly newspaper, newsletter, and specialty publications.

NBD PRODUCT-INFORMATION SYSTEMS BV

P.O. Box 23
7400 GA Deventer

CONTACT: F. R. Thedieck, Director

MISSION: To serve as intermediary between the producer and the building product user.

PRIMARY WORK: Publishes the *Netherlands Building Documentation* system.

SOURCE OF FINANCES: Product manufacturers using the service.

PUBLICATIONS: Building product index, specialty product information, and bulletins.

PHILIPS INTERNATIONAL BV, BUILDING DESIGN AND ENGINEERING DIVISION

Cederlaan 4
5616 SC Eindhoven

CONTACT: A. B. M. van der Plas

MISSION: To provide architectural and design services to users, as well as services relating to accommodations and communications.

PRIMARY WORK: Performs consultations in civil, mechanical, electrical engineering; project management; feasibility studies; managing, operating, and maintaining accommodations, and all related support functions related to those tasks; developing and stimulating an effective environmental and energy policy; and planning and implementing communications policy, systems, and facilties.

SOURCE OF FINANCES: Fees for services.

PUBLICATIONS: Professional journals.

PLANNING COMMITTEE FOR BUILDING RESEARCH OF THE NETHERLANDS ORGANIZATION FOR APPLIED SCIENTIFIC RESEARCH

P.O. Box 238
2600 AE Delft

MISSION: To serve as a laboratory and testing facility for the building community.

PRIMARY WORK: Performs research in building physics, environmental engineering, building materials, metals, corrosion, paints, surface technology, physics of explosions, structures, fire engineering, plastics and rubber, and risk analysis.

SOURCE OF FINANCES: Government, grants and contributions, and fees for research.

PUBLICATIONS: Journal, research reports, papers in professional journals, committee reports, and specialty publications.

RIW INSTITUTE FOR HOUSING RESEARCH

Berlageweg 1
2628 CR Delft

CONTACT: P. P. J. Houben, Director

MISSION: To provide and implement improved housing technology; Associated with the Delft Technical University.

PRIMARY WORK: Performs research in housing economics, housing quality, housing needs for special groups, and urban renewal.

SOURCE OF FINANCES: Delft Technical University and fees for research.

PUBLICATIONS: Articles in professional journals, research results, and specialty publications.

SOCIETY FOR THE STUDY OF PRECAST CONCRETE CONSTRUCTION

Raamweg 16
P.O. Box 85512
2508 CE 's-Gravenhage

CONTACT: F. Treffers, Managing Director

MISSION: To serve building community members with improved technologies and methods.

PRIMARY WORK: Performs research in materials including precast concrete and connection details and develops codes of practice for precast concrete.

SOURCE OF FINANCES: Annual member fees and research grants.

PUBLICATIONS: Research reports and published articles in the Dutch Concrete Institute periodicals.

THE NETHERLANDS WATERWORKS TESTING AND RESEARCH INSTITUTE (KIWA, LTD)

<u>Head Office and Testing Division</u>
Sir Winston Churchill Laan 273
P.O. Box 70
Rijswijk 2280 AB

<u>Research Division</u>
Groningenhaven 7
P.O. Box 1072
Nieuwegein 3430 BB

CONTACT: G. Martijn, Managing Director

MISSION: To serve as the central testing and research institute for The Netherlands waterworks.

PRIMARY WORK: Undertakes certification and inspection of products for waterworks, including water systems for buildings; and research for waterworks (hrydrology, catchment, purification, transport and distribution, water quality, corrosion).

SOURCE OF FINANCES: Fees for certification, contribution of waterworks association, charges for inspection, and research projects.

PUBLICATIONS: Professional journals, conference proceedings, research reports, and test specifications.

NEW ZEALAND

BUILDING RESEARCH ASSOCIATION OF NEW ZEALAND (BRANZ)

Private Bag
Porirua, Wellington

CONTACT: Peter K. Foster

MISSION: BRANZ is an industry-backed independent research and testing organization set up to acquire, apply and distribute knowledge about building that will benefit the industry in New Zealand, and, through it, the community at large.

PRIMARY WORK: Performs research and testing in energy conservation, moisture control, wind and earthquake engineering, materials, durability, fire engineering, and information systems; evaluates innovative systems and products; and provides training and advisory services.

SOURCE OF FINANCES: Levy on building construction, government, membership fees, testing and evaluation fees, and sales of publications.

PUBLICATIONS: Newsletter, research reports, technical papers, information bulletins, annual report, list of publications, appraisal certificates, selected building references, technical recommendations, program of work, and audiovisuals.

FOREST RESEARCH INSTITUTE, FOREST PRODUCTS DIVISION

Private Bag
Rotorua

CONTACT: A. J. McQuire, Director

MISSION: To provide the building community with improved production and use of forest products.

PRIMARY WORK: Research is performed in timber engineering, wood quality and preservation, timber drying, adhesives, and biotechnology.

SOURCE OF FINANCES: Government.

PUBLICATIONS: Journal, research reports, and publication list.

MINISTRY OF WORKS AND DEVELOPMENT CENTRAL LABORATORIES

P.O. Box 30845
Lower Hutt

CONTACT: J. H. Wood, Director

MISSION: To provide laboratory support for improving the design, construction, and maintenance of buildings.

PRIMARY WORK: Performs research in soil and rock mechanics, foundation engineering, structural engineering, materials, concrete and bitumen, hydraulic and aerodynamic modeling, and building sciences.

SOURCE OF FINANCES: Government and fees for research.

PUBLICATIONS: Research reports and list of publications.

NEW ZEALAND CONCRETE RESEARCH ASSOCIATION

13 Wall Place
Private Bag, Porirua

CONTACT: David P. Barnard, Director

MISSION: To provide members with technical services for improving their building practices.

PRIMARY WORK: Performs research in the characterization and performance of concrete and concrete application, and conducts training for design and construction practices.

SOURCE OF FINANCES: Cement industry and government.

PUBLICATIONS: Monthly journal and information bulletins.

NEW ZEALAND HEAVY ENGINEERING RESEARCH ASSOCIATION (HERA)

P.O. Box 76-134
Manukau City

CONTACT: Llew Richards

MISSION: HERA is the research arm of the New Zealand Heavy Engineering Industry. It is recognized as the national center of excellence for New Zealand engineering providing a foundation for the sound development of the industry.

PRIMARY WORK: Performs research on structural steel, structural design, welding techniques and technologies, market opportunities, and the industrial sector; and carries out an education and training function in each of these areas.

SOURCES OF FINANCE: Levy on sales of heavy steel, government contribution, membership fees, and sales of publications.

PUBLICATIONS: Monthly newsletter, quarterly bulletin, regular technical reports, annual reports, and selected overseas references.

ROAD RESEARCH UNIT, NATIONAL ROADS BOARD (RRU NRB)

P.O. Box 12 041
Wellington North

CONTACT: E. J. Burt, Chairman

MISSION: To foster an effective road transportation system through improved quality and efficient of the nation's road system.

PRIMARY WORK: Performs research on structural and bridge engineering, seismic analysis, foundation engineering, economics, road maintenance and management, road and bridge building materials, concrete, transportation safety, and environmental engineering.

SOURCE OF FINANCES: Government and grants and contracts awarded to appropriate agencies.

PUBLICATIONS: Quarterly newsletter, technical recommendations, research bulletins, and specialty publications.

AUCKLAND UNIVERSITY, SCHOOL OF ARCHITECTURE

22 Symonds Street
Auckland 1

CONTACT: Allan Wild, Director

MISSION: To provide education and research laboratory support for improving building practices.

PRIMARY WORK: Undertakes research in energy conservation, building acoustics, economics, and computer-aided design.

SOURCE OF FINANCES: University, grants, and fees for research.

PUBLICATIONS: Technical standards, research reports, and bibliographies.

CANTERBURY UNIVERSITY, DEPARTMENT OF CIVIL ENGINEERING

Private Bag
Christ Church

CONTACT: R. Park, Department Head

MISSION: To provide education and building sciences to advance building practices.

PRIMARY WORK: Performs research on properties of materials and structural, earthquake, and foundation engineering.

SOURCE OF FINANCES: University and external agencies in New Zealand.

PUBLICATIONS: Research reports, theses, list of publications, papers in conference, proceedings, and technical journals.

VICTORIA UNIVERSITY OF WELLINGTON, SCHOOL OF ARCHITECTURE

Private Bag
Wellington

CONTACT: Wendy Light, Chairperson

MISSION: To provide education and architectural and building sciences research to improve building practices.

PRIMARY WORK: Performs research on architecture, lighting, acoustics, building construction, building failures, and timber structures.

SOURCE OF FINANCES: University and fees for research.

PUBLICATIONS: Research reports.

NIGERIA

IMO STATE HOUSING CORPORATION

Private Mail Bag 1224
Owerri, Imo State

CONTACT: P. C. Chikwendu, General Manager

MISSION: To implement government housing schemes for development and construction of adequate housing for the population.

PRIMARY WORK: Manages the housing implementation program, maintains appropriate building standards, encourages development of indigenous materials, and urges improvements in the quality of housing.

SOURCE OF FINANCES: Commercial banks.

PUBLICATIONS: Quarterly journal.

NORWAY

ACOUSTICS LABORATORY

N 7034
Trondheim NTH

CONTACT: Jens Trampe Broch, Director

MISSION: To serve as the national center for acoustic noise research.

PRIMARY WORK: Performs research in electro-acoustics, noise and vibrations, psycho-physio-acoustics, and architectural acoustics.

SOURCE OF FINANCES: Government and industry.

PUBLICATIONS: Research reports and annual journal.

**NATIONAL INSTITUTE OF TECHNOLOGY,
DEPARTMENT OF BUILDING AND CONSTRUCTION (STI)**

Akersvn 24c
P.O. Box 8116
Oslo 1

CONTACT: Alf Hernes, Administrative Director

MISSION: To serve as national center for development, information, education, and consulting for the building industry.

PRIMARY WORK: Procures research in civil engineering, construction, water supply and sewerage, materials, concrete, timber, paints, surface treatment; and performs laboratory work and tests on concrete, safety flat glass for building (hammer test, ball test, shot bag test, bullet resistant tests).

SOURCE OF FINANCES: Government and fees for research.

PUBLICATIONS: Conference proceedings, specialty publications, textbooks, and means of instruction.

NORWEGIAN BUILDING CENTER

P.O. Box 1575
Vika, Oslo 1

CONTACT: Endre Hoflandsdal, Managing Director

MISSION: To provide building association members with centralized information for improving their knowledge about building technology.

PRIMARY WORK: Maintains an information department, coordinates exhibits, and develops information and delivery systems.

SOURCE OF FINANCES: Sales of publications and fees for services.

PUBLICATIONS: Data sheets, building regulations, and codes of practice; materials listings.

NORWEGIAN BUILDING RESEARCH INSTITUTE (NBI)

P.O. Box 123
0314, Oslo 3

CONTACT: Aage Hallquist, Director

MISSION: To serve as national building research laboratory to advance technology for improved building practices.

PRIMARY WORK: Performs research in energy conservation, alternate energy resources, lighting, acoustics, HVAC, building materials, construction engineering, building processes, information management, and economics.

SOURCE OF FINANCES: Sponsored research, fees from industry, sales of publications, and government.

PUBLICATIONS: Semiannual data sheets, research reports, handbooks, design manuals, testing methods, list of publications, and specialty publications.

NORWEGIAN GEOTECHNICAL INSTITUTE (NGI)

Sognsveien 72
OSLO 8

CONTACT: Kaare Hoeg, Director

MISSION: To advance the state of knowledge of geotechnical engineering.

PRIMARY WORK: Performs research in dam and rock engineering, foundation engineering, and marine geotechnology.

SOURCE OF FINANCES: Government, grants, and contracted research.

PUBLICATIONS: Research reports, publication series, conference papers, articles in professional journals, and specialty publications.

NORWEGIAN INSTITUTE OF WOOD TECHNOLOGY (NTI)

Forskningsveien 3b
Blindern
Oslo 3

CONTACT: Rolf Birkeland, Managing Director

MISSION: To supply the wood industry with advanced timber technology.

PRIMARY WORK: Performs research in wood technology, surface treatment and gluing, and structural engineering.

SOURCE OF FINANCES: Industry fees, grants, and fees for contracted research.

PUBLICATIONS: Research reports and specialty reports.

SINTEF DEPARTMENT OF ARCHITECTURE AND BUILDING TECHNOLOGY

Alfred Getz vei 3
7034 Trondheim-Nth

CONTACT: Sigmund Asnervik, Manager

MISSION: To serve as the university's research and development center and to provide the link to the building community.

PRIMARY WORK: Performs research in solar energy, daylighting, rehabilitation. energy utilization and conservation, computer-aided design, and environmental engineering.

SOURCE OF FINANCES: Government, fees for research, and industry grants.

PUBLICATIONS: Research reports and specialty publications.

SINTEF NORWEGIAN FIRE RESEARCH LABORATORY (NBL)

N 7034
Trondheim-NTH

CONTACT: Kjell Schmidt Pedersen, Director

MISSION: To serve as a center for fire research and fire testing.

PRIMARY WORK: Performs research in fire testing, fire control, and evaluation of materials; classifies building materials, constructions, and fire equipment; and offers advice and consultancy.

SOURCE OF FINANCES: Government, industry, and fees for research from research councils.

PUBLICATIONS: Research reports and annual report.

PAKISTAN

BUILDING RESEARCH STATION (BRS)

F 40
Sind Industrial Trading Estate
Hub River Road
Karachi 28

CONTACT: M. Sulaiman, Director

MISSION: To serve as research and testing center for providing technology to meet the need for affordable and adequate low-cost housing.

PRIMARY WORK: Performs research in soils engineering, building materials and components, utilization of natural resources, and maintenance; develops information systems.

SOURCE OF FINANCES: Government.

PUBLICATIONS: Technical recommendations, research reports, and annual report.

BUILDING RESEARCH STATION LAHORE (BRS LAHORE)

P.O. Box 1230
Near New University Campus
Lahore

CONTACT: Qamar-ur-rehmar Mufti, Director

MISSION: To provide improved technology for better use of indigenous materials.

PRIMARY WORK: Performs research in building construction, strength of materials, soils and foundation engineering, concrete and aggregate, and lime and gypsum.

SOURCE OF FINANCES: Government.

PUBLICATIONS: Research reports, articles in professional press, and bulletins.

PHILIPPINES

UNIVERSITY OF THE PHILIPPINES, BUILDING RESEARCH SERVICE, NATIONAL ENGINEERING CENTER

102 National Engineering Center Building
University of the Philippines
Diliman, Quezon City

CONTACT: Geronimo V. Mahahan, Director

MISSION: To serve as a research and development center for advancing the understanding of material properties, their production, and use.

PRIMARY WORK: Materials testing and geotechnical services; consultancy and technical assistance; and research and development on building materials, construction systems, community development, urban ecology, and low cost energy source and conservation; publications.

SOURCE OF FINANCES: University, research contracts, and grants.

PUBLICATIONS: Newsletter, research, reports, bibliography, guidelines, and specialty publications.

FOREST PRODUCTS RESEARCH AND DEVELOPMENT INSTITUTE (FPRDI), DEPARTMENT OF SCIENCE AND TECHNOLOGY (DOST)

College Laguna 3720

CONTACT: Florentino O. Tesoro, Director

MISSION: The institute is committed to the attainment of national development goals by generating, through research and development, technical information and technologies in the utilization of forest products.

PRIMARY WORK: The Housing Materials Division, one of three technical divisions of the institute, conducts research on wood anatomy and dendrology, timber physics and mechanics, preservation and protection, composite panels, and timber engineering.

SOURCE OF FINANCES: Government, grants in aid, and research contracts.

PUBLICATIONS: *FPRDI Journal* and *Forest Products Technoflow*.

PHILIPPINE COUNCIL FOR INDUSTRY AND ENERGY RESEARCH AND DEVELOPMENT (PCIERD)

Rm. 513, 5th Floor
Ortigas Building
Ortigas Avenue
Pasig, MN

CONTACT: Benjamin T. Damian, Executive Director

MISSION: To serve as the central agency in the planning, monitoring, and promotion of scientific and technological research for applications in the industry, energy utilities, and infrastructure sectors.

PRIMARY WORK: Research, development, planning, monitoring.

SOURCE OF FINANCES: Government and foriegn grants/assistance.

PUBLICATIONS: Quarterly newsletters, annual reports, and technical brochures.

UNITED ARCHITECTS OF THE PHILIPPINES

Cultural Center of the Philippines
Roxas Boulevard, Metro Manila

CONTACT: Dorothy Lucasan, Administrative Officer

MISSION: Professional organization of registered architects accredited by the Philippine government established to promote highest standard of ethical conduct in the practice, uplift standards of architectural education, and disseminate information in ecology, technology and environmental design.

FINANCES: Annual dues, grants, and donations.

PUBLICATIONS: Quarterly newsletter and specialty publications.

POLAND

BUILDING RESEARCH INSTITUTE (ITB)

ul Filtrowa 1
00-950 Warsaw

CONTACT: Marian Weglarz, Director

MISSION: To improve technology in construction and building materials and to provide for standardization and application of new building materials.

PRIMARY WORK: Performs research in structural engineering, construction technologies, fire engineering, building materials, protection of building objects, physics and acoustics in buildings, finishing techniques, geotechnics, and building objects in mining areas.

SOURCE OF FINANCES: Government and sale of services within contracts.

PUBLICATIONS: Research reports, quarterly and annual journals, and technical recommendations.

INSTITUTE FOR BUILDING MECHANIZATION AND ROCK MINING

ul Racjonalizacji 6/8
02-673 Warsaw

CONTACT: E. Budny, Director General

MISSION: To undertake research in the areas of building mechanization, earth-moving and construction machinery, and rock mining.

PRIMARY WORK: Preparing basic performance data for production of new machinery and advanced technologies in new machines, providing services in testing of machines and equipment, and undertaking projects in building mechanization.

SOURCES OF FINANCE: National central research and development programs and individual contracts with construction and mining equipment manufacturers.

PUBLICATIONS: *Problems of Building Mechanization,* quarterly inserts in *Mechanical Review* and *Building Review,* brochures, instructions, and recommendations.

MAIN CENTER FOR BUILDING INFORMATION

ul Senatorska 27
00-950 Warsaw

CONTACT: Henryk Watcerz, Manager

MISSION: To serve as the main center for collecting and disseminating technical information for the building community.

PRIMARY WORK: Provides services to automate information processing, develops information systems, and hosts exhibitions.

SOURCE OF FINANCES: Sale of publications, and subsidies.

PUBLICATIONS: Research reports, bulletins, building abstracts, and catalogues.

RESEARCH AND DESIGN CENTER FOR INDUSTRIAL BUILDING

ul Parkingowa 1
00-518 Warsaw

CONTACT: A. Bratkowski, Director General

MISSION: To furnish research and design services for the building community.

PRIMARY WORK: Provides services in research and design of projects for industrial and public buildings, including district heating, refrigerating engineering, and environmental protection; computer-aided design systems.

SOURCE OF FINANCES: Contract work.

PUBLICATIONS: Bimonthly journal and periodic publications.

RESEARCH INSTITUTE FOR ENVIRONMENTAL DEVELOPMENT (RIED)

ul Krzywickiego 9
02-078 Warsaw

CONTACT: Pawet Btaszczyk, Director General

MISSION: To provide planning theory and practice for human settlement ecosystems.

PRIMARY WORK: Performs work in diagnostics, maintenance, human settlement technology, environment, and housing.

SOURCE OF FINANCES: Government.

PUBLICATIONS: Quarterly and monthly journals and bulletins.

RESEARCH INSTITUTE OF MINERAL BUILDING MATERIALS (IMMB)

ul Oswiecimska 21
45-641 Opole

CONTACT: Bronistaw Werynski, General Director

MISSION: To serve the cement, lime, and gypsum manufacturers with advanced technology.

PRIMARY WORK: Performs research on cement, lime, and gypsum; energy savings in mineral building materials; automation of industrial control systems; environmental engineering; and industrial waste utilization.

SOURCE OF FINANCES: Government and cement, lime, and gypsum manufacturers.

PUBLICATIONS: Monthly journal and research reports.

ROAD AND BRIDGE RESEARCH INSTITUTE (IBDM)

ul Stalingradzka 40
03-301 Warsaw

CONTACT: Mieczystaw Rybak, Managing Director

MISSION: To introduce new structural techniques, equipment, and design and construction principles to improve road and bridge building practices.

PRIMARY WORK: Performs work in materials, civil engineering, geotechnical engineering, construction, and road equipment.

SOURCE OF FINANCES: Government.

PUBLICATIONS: Quarterly bulletin, newsletter, information series, and specialty publications.

WOOD TECHNOLOGY INSTITUTE (ITD)

ul Winiarska 1
60-654 Poznan

CONTACT: R. Babicki, Director

MISSION: To advance wood technology for use by the wood product manufacturers.

PRIMARY WORK: Performs research in the chemistry, preservation, processing, treatment, gluing, applications, insulation, and economics of wood.

SOURCE OF FINANCES: Government and wood industry.

PUBLICATIONS: Quarterly journal.

PORTUGAL

FOREST PRODUCTS INSTITUTE (IPF)

Rua Filipe Folque
10-J
1000 Lisbon

CONTACT: Joao Manuel Alves Soares, President

MISSION: To provide improved technology for advancing cork, resinous products, and wood and its derivatives.

PRIMARY WORK: Performs laboratory support, test methods, economic studies, and quality control.

SOURCE OF FINANCES: Levy from forest products industry.

PUBLICATIONS: Monthly and quarterly bulletins, annual report, and specialty publication.

NATIONAL LABORATORY OF CIVIL ENGINEERING (LNEC)

Avenida do Brasil 101
1799 Lisbon Codex

CONTACT: Artur Ravara, Director

MISSION: National research laboratory to advance building practices.

PRIMARY WORK: Performs research in building materials, concrete, plastics, structures, timber, geotechnical engineering, foundations, hydraulics, acoustics, lighting, economics, HVAC, fire protection, and transportation.

SOURCE OF FINANCES: Government.

PUBLICATIONS: Monthly bulletins, research reports, technical recommendations, Agrement documents, test results, specifications, standards, proceedings, and specialty publications.

SOUTH OF PORTUGAL'S BUILDING AND PUBLIC WORKS ASSOCIATION (AECOPS)

Rua Antonia Enes 9-5
1000 Lisbon

CONTACT: Jose J. Tomas Gomes, General Director

MISSION: To furnish technical and administrative support to the building and public works contractors.

PRIMARY WORK: Provides services in civil engineering applications, construction materials, regulations, economics, work agreements, and construction practices.

SOURCE OF FINANCES: Membership fees.

PUBLICATIONS: Monthly journal, research reports, catalogue, magazines, and specialty publications.

TECHNICAL ASSOCIATION OF THE CEMENT INDUSTRY (ATIC)

Av 5 de Outubro
54-2, Dt
1000 Lisbon

CONTACT: Manuel Lourenco Antunes, Director

MISSION: To represent the cement industry with knowledge for improving the manufacturing and application process.

PRIMARY WORK: Perform works in development of cement and mortars, information systems, and delivery methods.

SOURCE OF FINANCES: Member contributions.

PUBLICATIONS: Technical results and specialty publications.

PORTO UNIVERSITY FACULTY OF ENGINEERING, BUILDING CONSTRUCTION SECTION

Rua dos Bragas
4099 Porto Codex

CONTACT: Vitor Abrantes, Director

MISSION: To educate and serve the building community with improved acoustics and hygrometric technology.

PRIMARY WORK: Performs research in acoustics, hygrometrics, fire, lighting, economics, and productivity and testing services.

SOURCE OF FINANCES: Government and fees for research.

PUBLICATIONS: Bimonthly bulletins and specialty publications.

ROMANIA

BUILDING DESIGN INSTITUTE FOR TYPIFIED CONSTRUCTION (IPCT)

Code 70132 Bucharest
Str. Tudor Arghezi 21

CONTACT: Petru Vernescu, Engineer, Director

MISSION: To carry out projects for industrial, agricultural, and civil buildings and to establish solutions for construction with a high degree of industrialization and superior performance regarding functionality and durability.

PRIMARY WORK: Performs typified design in the field of industrial buildings, silos, water towers, water tanks, apartment blocks, office buildings, social and cultural buildings, thermal power stations, and solar power stations. Also performs computer programs for structural engineering (especially for seismic loads).

SOURCE OF FINANCES: Contracted design research and services.

PUBLICATIONS: Quarterly journal and design catalogues.

NATIONAL RESEARCH INSTITUTE FOR CONSTRUCTION AND CONSTRUCTION ECONOMICS (INCERC)

sos Pantelimer 266
Bucharest 2

CONTACT: Valeriu Cristiscu, General Director

MISSION: To promote new technologies and testing methods and to implement technologies into the market place.

PRIMARY WORK: Undertakes research in reinforced and prestressed concrete, steel structures, earthquake and geotechnical engineering, acoustics, fire engineering, building physics, and mechanization of buildings.

SOURCE OF FINANCES: Government, research contracts, and fees from concrete industry.

PUBLICATIONS: Monthly journal, technical standards, research reports, and specialty publications.

CENTRAL BUILDING RESEARCH INSTITUTE, DEPARTMENT OF CLUJ-NAPOCA

Calea Foresti 117
3400 Cluj-Napoca

CONTACT: Adrian Ioani, Department Director

MISSION: To advance civil works technology and perform material testing.

PRIMARY WORK: Provides research in civil engineering, structural mechanics, lightweight concrete, and building technology.

SOURCE OF FINANCES: Government and fees for contracts.

PUBLICATIONS: Annual information bulletin, research results, and specialty publications.

DOCUMENTATION INFORMATION OFFICE FOR BUILDING, ARCHITECTURE, AND TOWN PLANNING (CDCAS)

Boul 1848 no 10
CP 1-139
Bucharest cod 70058

CONTACT: Aurelia Dobrescu, Chief

MISSION: To provide technical and scientific information for the building community.

PRIMARY WORK: Performs work in documentation and information systems, architecture and town planning, construction technologies, industrialized systems, and building equipment.

SOURCE OF FINANCES: Sales of publications.

PUBLICATIONS: Bimonthly building product and desIgn information, bulletins and bibliographies on building materials and building technology, document surveys, and technical abstracts.

RESEARCH AND DESIGN INSTITUTE FOR PHYSICAL PLANNING, HOUSING, AND MUNICIPAL ENGINEERING (ISLGC)

Str. Snagov 53-55
70136 Bucharest

CONTACT: Adrian Alecandrescu, Director

MISSION: To provide technologies for improving urban and rural planning.

PRIMARY WORK: Performs work in town and country planning, national housing programs, and economics.

SOURCE OF FINANCES: Government and fees for services.

PUBLICATIONS: Results of studies, guidelines, and specialty publications.

SINGAPORE

CONSTRUCTION INDUSTRY DEVELOPMENT BOARD

133 Cecil Street 09-01 Keck Seng Tower
Singapore 0106

CONTACT: Chee Yoong Ng, General Manager

MISSION: To promote improvements in the planning and coordinating of the development of the construction industry.

PRIMARY WORK: Performs research and services to facilitate the mechanization and efficiency of the construction industry and provides training facilities.

SOURCE OF FINANCES: Government.

PUBLICATIONS: Quarterly review, monthly construction market report, directory of construction equipment, and prefabricated concrete components.

**SINGAPORE NATIONAL UNIVERSITY,
FACULTY OF ARCHITECTURE AND BUILDINGS**

Kent Ridge Campus
Singapore 0511

CONTACT: Registrar

MISSION: Education and research to improve building practices.

PRIMARY WORK: Undertakes research in architectural design, planning, acoustics, lighting, energy conservation and renewal, structures, thermal behavior, economics, and management.

SOURCE OF FINANCES: Government.

PUBLICATIONS: Research reports and articles in professional journals.

SOUTH AFRICA

AGREMENT BOARD OF SOUTH AFRICA

P.O. Box 395
Pretoria 0001

CONTACT: C. J. Schlotfeldt, General Manager

MISSION: To perform Agrement certification procedures for more effective use of resources and improved product quality.

PRIMARY WORK: Evaluates fitness-for-purpose of building innovations, establishes performance criteria and test methods, and certifies and monitors quality.

SOURCE OF FINANCES: Fees for services and government.

PUBLICATIONS: Performance criteria, certifications, explanatory/promotional documents, indexes of certifications and publications, and annual report.

BUILDING INDUSTRIES FEDERATION (SOUTH AFRICA) (BIFSA)

Federated Insurance House
1 De Villiers Street
11359 Johannesburg

CONTACT: Robert Zylstra, Executive Director

MISSION: To promote good building and safety practices and labor relations.

PRIMARY WORK: Performs work in economics, education and training, materials, housing, safety, and information.

SOURCE OF FINANCES: Fees for services and government.

PUBLICATIONS: Manuals, annual report, training documents, and guidelines.

NATIONAL BUILDING RESEARCH INSTITUTE (NBRI)

P.O. Box 395
Pretoria 0001

CONTACT: John Morris, Chief Director

MISSION: National research laboratory to improve technology and to solve national problems with an emphasis on housing.

PRIMARY WORK: Performs research on housing, building services, energy consumption and conservation, environmental engineering, lighting, acoustics, natural ventilation, Agrement and building regulations, fire engineering, materials, concrete, paints, structural and geotechnical engineering, economics, and information delivery.

SOURCE OF FINANCES: Government and contracted research.

PUBLICATIONS: Research reports, proceedings, abstracts, information sheets, technical notes, articles in professional journals, and list of publications.

NATIONAL INSTITUTE FOR TRANSPORT AND ROAD RESEARCH (NITRR)

P.O. Box 395
Pretoria 0001

CONTACT: R. N. Walker, Chief Director

MISSION: To undertake research and development and to promote their implementation so as to make the transportation of people and goods in southern Africa more economical, efficient, safe, and socially responsible.

PRIMARY WORK: Performs research on design, construction, and maintenance of roads and bridges and provides testing services; conducts research into road and vehicle safety; researches and provides data on the following aspects of transportation: planning and development, statistics, economics, and marketing.

SOURCE OF FINANCES: Government.

PUBLICATIONS: Research reports, bulletins, various series, digest, technical recommendations, technical methods, abstracts, urban transport guidelines, and specialty publications.

PORTLAND CEMENT INSTITUTE (PCI)

PO Box 168
1685 Halfway House

CONTACT: S. W. Norton, Executive Director

MISSION: To provide improved technology on the properties and applications of cement and concrete.

PRIMARY WORK: Performs services in disseminating technical information, technical advisory service, materials testing, and training.

SOURCE OF FINANCES: Cement producers.

PUBLICATIONS: Research reports, list of publications, and specialty publications.

SPAIN

AGROMAN EMPRESA CONSTRUCTORA S.A., DIVISION OF RESEARCH AND METHODS

Raimundo Fernandez Villaverde 43
Madrid 28003

CONTACT: Jose Maris Aguirre, Chairman

MISSION: To provide technology for improved civil and industrial construction.

PRIMARY WORK: Performs research on construction development, materials and machines, and construction methods.

SOURCE OF FINANCES: Agroman Empresa Constructora S.A.

PUBLICATIONS: Research review and annual almanac.

INSTITUTE OF TECHNOLOGY OF THE CONSTRUCTION OF CATALUNA

Bon Pastor 5
08021 Barcelona 21

CONTACT: Josep Maria Valeri, Director

MISSION: To provide technology to improve building and civil engineering practices.

PRIMARY WORK: Performs work in development and adaption of building and civil engineering technology, training, documentation services, construction, economics, construction quality, regulations, construction data base, and applications programs.

SOURCE OF FINANCES: Government, fees for services, sales of publications, and dictionaries.

PUBLICATIONS: Quarterly journal, technical standards, research recommendations, and research reports.

CANTABRIA UNIVERSITY, SCHOOL OF CIVIL ENGINEERING

Av de los Castros s/n
Santander

CONTACT: Federico Gutierrez-Solana Salcedo, Director

MISSION: To provide education and technical training to improve building practices.

PRIMARY WORK: Performs research on geotechnical engineering, materials, concrete, energy conservation, statistics, economics, hydraulics, structures design, and mechanics.

SOURCE OF FINANCES: Government.

PUBLICATIONS: Research reports.

BUILDING AND CEMENT INSTITUTE EDUARDO TORROJA

Serrano Galvache
s/PO Box 19.002-28080
Madrid

CONTACT: Juan Murcia, Director

MISSION: To provide new knowledge in materials and structures.

PRIMARY WORK: Performs research on cement manufacturing, energy conservation, concrete durability, additives, industrialization techniques, and prestressed concrete.

SOURCE OF FINANCES: Grants.

PUBLICATIONS: Journals, monographs, and handbooks.

TECHNICAL INSTITUTE OF MATERIALS AND CONSTRUCTIONS (INTEMAC)

Monte Esquinza 30
28010 Madrid

CONTACT: Jose Calavera, General Director

MISSION: To provide methods to improve the quality control of construction materials.

PRIMARY WORK: Performs work in building pathology, materials, soil mechanics, geotechnical engineering, training, and testing.

SOURCE OF FINANCES: Fees for work.

PUBLICATIONS: Research reports.

ROAD RESEARCH CENTRE

Autovia de Colmenar P.K. 18,2
El Goloso
28049 Madrid

CONTACT: Jose L. Elvira, Director

MISSION: To advance the state of knowledge of roads and transportation engineering for improved applications.

PRIMARY WORK: Performs applied research on traffic and transportation, road construction, design, pavement materials, real-scale test tracks, and maintenance and management systems.

SOURCE OF FINANCES: Government and fees for research.

PUBLICATIONS: Bimonthly bulletin (CEDEX), technical standards, and research books.

MILITARY ENGINEERING LABORATORY

Serrano Jover
2 Madrid 8

CONTACT: D. Jose-Luis Cabanes Torrente, Director

MISSION: To perform testing services for the Ministry of Defense.

PRIMARY WORK: Provides services in soil mechanics, military equipment (except weapons), and materials.

SOURCE OF FINANCES: Government.

PUBLICATIONS: Research reports and technical reports.

SWEDEN

BUILDING STANDARDS INSTITUTION (BST)

Drottning Kristinas vag 73
114 28 Stockholm

CONTACT: Goran Stensgard, Director

MISSION: To develop building standards through voluntary standards committees.

PRIMARY WORK: Performs work in developing practices for building and civil engineering applications, general test methods, materials, components, modular coordination, and tolerances; and publishes and sells information.

SOURCE OF FINANCES: Government, members, and sales of publications.

PUBLICATIONS: Standards, handbooks, and annual report.

DEPARTMENT OF WORK SCIENCE

KTH, S 100 44 Stockholm

CONTACT: Ulf Ulfvarson, Director

MISSION: To provide improved methods for the labor work force in construction and other industries.

PRIMARY WORK: Performs research in ergonomics and industrial hygiene.

SOURCE OF FINANCES: Government and worker funds.

PUBLICATIONS: Research reports.

LUND INSTITUTE OF TECHNOLOGY, DEPARTMENT OF STRUCTURAL ENGINEERING

Box 725
S 220 07 Lund

CONTACT: Per Christiansson, Associate Professor
Lars Ostlund, Emeritus Professor Department of Structural Engineering

MISSION: To educate and advance computer applications for civil engineering and to develop new knowledge about loads on structures and quality assurance of building elements.

PRIMARY WORK: Performs research on knowledge-based systems, integrated systems design, design of structures, loads on building components, risk analysis, and training.

SOURCE OF FINANCES: University, fees for services, and industry grants.

PUBLICATIONS: Research reports, manuals, conference papers, and information sheets.

NATIONAL BOARD OF PHYSICAL PLANNING AND BUILDING

Vasterbroplan
Box 12513
S 102 29 Stockholm,

CONTACT: Dennart Holm, Director General

MISSION: Supervises national planning and building projects and issues permits.

PRIMARY WORK: Codes and guidelines for building design and construction, and for rural and urban planning.

SOURCE OF FINANCES: Government.

PUBLICATIONS: Information publications, bimonthly journal, and statutory publications.

NATIONAL SWEDISH INSTITUTE FOR BUILDING RESEARCH (SIB)

Box 785
S-801 29 Gavle

CONTACT: Nils Antoni, Director

MISSION: National research laboratory that focuses on improving technologies to reduce housing shortages, improve the quality of life, and produce better measurement and test methods.

PRIMARY WORK: Performs research on energy conservation, materials, building climatology, water supply and drainage, building technology, housing, production, construction, land use policies, economics, information systems, and building planning.

SOURCE OF FINANCES: Government.

PUBLICATIONS: Research results, lists of publications, and specialty publications.

NATIONAL SWEDISH ROAD AND TRAFFIC RESEARCH INSTITUTE (VTI)

S 581 01 Linkoping

CONTACT: Hans Sandebring, Director General

MISSION: To undertake research and development to improve design and construction practices and to communicate research results.

PRIMARY WORK: Performs work in foundations engineering, civil engineering, construction, ergonomics, traffic engineering, measurement methods, and information systems.

SOURCE OF FINANCES: Government.

PUBLICATIONS: Bimonthly journal, research reports, bibliographies, and annual reports.

THE SWEDISH NATIONAL TESTING INSTITUTE (SP)

Brinellgatan 4
Box 857
501 15 Boras

CONTACT: Claes Bankvall, Director General

MISSION: National authority for testing, inspection, and metrology.

PRIMARY WORK: Testing, research and development, and advisory work in building technology; building physics; fire technology; energy and environmental technology; materials and mechanics; physics and electrotechnics; and metrology.

SOURCE OF FINANCES: Fees for testing, grants for research and advisory work from private organizations, research boards, and government.

PUBLICATIONS: Technical report series, test methods, and technical information series.

ROYAL INSTITUTE OF TECHNOLOGY, DIVISION OF BUILDING TECHNOLOGY

Brinevellvagen 34
S 100 44 Stockholm

CONTACT: Ingemar Hoglund, Director

MISSION: To provide education and research in building physics and design.

PRIMARY WORK: Performs research in building physics, thermal engineering, environmental conditions, energy conservation, sealants, and passive solar systems.

SOURCE OF FINANCES: Government.

PUBLICATIONS: Research reports, list of publications, and specialty publications.

SIPOREX CENTRAL LABORATORY (SCL)

Stromgatan 11
Box 21053
200 21 Malmo

CONTACT: Oystein Kalvenes, Director

MISSION: Corporate laboratory for steam-cured concrete.

PRIMARY WORK: Performs research on building materials, concrete, corrosion, structural properties, surface treatment, and process development.

SOURCE OF FINANCES: Siporex Corporation.

PUBLICATIONS: Research reports to Siporex licensees.

SWEDISH BUILDING CENTRE

P.O. Box 7853
S 103 99 Stockholm

CONTACT: Henry Karlsson, Technical Director
Jan Lindgren, Managing Director

MISSION: To improve building practices by processing and disseminating building technology information including creating tools (classification systems with applications) for structuring information.

PRIMARY WORK: Perform works in document coordination, information systems, and publishing; supports permanent exhibition; prepares and publishes *National Building Specification (Master Specification)*.

SOURCE OF FINANCES: Sale of information and grants.

PUBLICATIONS: Technical studies, catalogues, building products facts, indexes, recommendations, *Master Specification*, handbooks, and manuals.

SWEDISH CEMENT AND CONCRETE INSTITUTE (CBI)

Drottning Kristinas vag 26
S-100 44 Stockholm

CONTACT: Bo G. Hellers, Director

MISSION: To improve concrete technology for better building practices and wider implementation and to work in joint cooperation with other concrete research organizations.

PRIMARY WORK: Performs work in material characterization, testing, and training and advice.

SOURCE OF FINANCES: Government, concrete industry, and fees for courses and services.

PUBLICATIONS: Research reports, information and task reports, annual report, and list of publications.

SWEDEN CORROSION INSTITUTE (SCI)

Drottning Kristinas vag 45
47 D and 48, Stockholm

CONTACT: Einar Mattsson, Director

MISSION: To perform research and disseminate new technologies in corrosion for reducing building failures.

PRIMARY WORK: Provides work in materials, anticorrosion methods, coatings, paint, surface coatings, and information systems.

SOURCE OF FINANCES: Government, grants, fees for services, and sale of information.

PUBLICATIONS: Journals, newsletters, technical standards, research results, and list of publications.

SWEDISH COUNCIL FOR BUILDING RESEARCH (BFR)

Sankt Goranagatan 66
S 112 33 Stockholm

CONTACT: Rune Olsson, Director General

MISSION: The Swedish Council for Building Research (BFR) is a sectorial research agency under the auspices of the Ministry of Housing and Physical Planning. It is responsible for the initiation, coordination, funding, and evaluation of R&D in the building and housing sector. Work is entrusted to universities, institutes of technology, socialized research institutes, public authorities, private companies, and individual researchers.

PRIMARY WORK: Supports R&D in the fields of urban design and management, building technology and energy conservation, and energy end use in buildings; distributes information about R&D results.

SOURCE OF FINANCES: Government.

PUBLICATIONS: Research reports, research programs and activity plans, *Synopses* publication, abstracts publication, and newsletter and magazine.

SWEDISH ENVIRONMENTAL RESEARCH INSTITUTE (IVL)

Halsingegatan 43, PO Box 21060
S 100 31 Stockholm

CONTACT: Allan Gustafsson, Managing Director

MISSION: To develop technology to reduce industrial air and water pollution.

PRIMARY WORK: Provides work in solid waste conversion, measuring air pollution, energy conservation, training, and testing.

SOURCE OF FINANCES: Government and grants.

PUBLICATIONS: Quarterly journal and research reports.

SWEDISH GEOTECHNICAL INSTITUTE (SGI)

581 01 Linkoping

CONTACT: Jan Hartien, Director General

MISSION: Seeks to achieve better techniques, safety and economy by the correct application of geotechnical knowledge in the building process.

PRIMARY WORK: Deals with geotechnical research, information, and consultancy; development of techniques for soil improvement and foundation engineering; environmental and energy geotechnics; design and development of field and laboratory equipment; and provides the Swedish central geotechnical literature service and computerized system.

SOURCE OF FINANCES: Government.

PUBLICATIONS: Research reports and monthly list of library accessions.

SWEDISH INSTITUTE OF BUILDING DOCUMENTATION, BYGGDOK

Halsingegatan 49
S-113 31 Stockholm

CONTACT: Bengt Eresund, Managing Director

MISSION: Central organization providing the building community with Scandanavian building technology information and documentation services.

PRIMARY WORK: Performs documentation services, literature searches, data base host and producer, information system, information broker, and delivery methods.

SOURCE OF FINANCES: Government and fees for services.

PUBLICATIONS: Technical standards, recommendations, research reports, abstract journals, general publications, and searches.

SWEDISH INSTITUTE FOR STEEL CONSTRUCTION (SBI)

Drottning Kristinal vag 48
S 114 28 Stockholm

CONTACT: P. O. Thomasson, Director

MISSION: To develop advances in research and development for steel construction.

PRIMARY WORK: Performs work in material properties, steel frame design and construction, fire protection, corrosion, bridges and buildings, regulations, codes and standards, information systems, and advisory services.

SOURCES OF FINANCES: Steel industry and research contracts.

PUBLICATIONS: Handbooks, manuals, research reports, and specialty publications.

SWITZERLAND

CONCRETE ROADS, LTD

Linderstrasse 10
CH-5103 Wildegg

CONTACT: W. Wilk, Director

MISSION: To provide expertise on concrete roads, airfields, industrial pavements, and soil stabilization.

PRIMARY WORK: Research, studies, and assistance to project designers in the above-mentioned fields; supervision of jobs onsite; and maintenance, testing, and road structural design.

SOURCE OF FINANCES: Contributions from association of cement factories and fees for services.

PUBLICATIONS: Bulletin, professional journals, and specialty publications.

SWISS FEDERAL INSTITUTE OF TECHNOLOGY, DEPARTMENT OF CIVIL ENGINEERING, LAUSANNE (ICOM)

GCB (Ecublens)
CH-1015 Lausanne

CONTACT: John C. Badoux, Director

MISSION: To provide education and to develop advances in civil engineering technology.

PRIMARY WORK: Performs work in steel structures, composite steel-concrete construction, civil engineering, training, and construction.

SOURCE OF FINANCES: University, government, and fees for services.

PUBLICATIONS: Monographs, scientific papers in journals, Ph.D. theses, list of publications, and specialty publications.

SWISS FEDERAL INSTITUTE OF TECHNOLOGY, RESEARCH INSTITUTE FOR THE BUILT ENVIRONMENT, LAUSANNE (IREC)

14 av Eglise-Anglaise
1006 Lausanne

CONTACT: Joseph Csillaghy, Director

MISSION: To provide education and advance technology in economics and social behavior concerning buildings.

PRIMARY WORK: Performs work in human behavior, settlements, urban development, economics, sociological aspects, and construction and management.

SOURCE OF FINANCES: University, government, and fees for research.

PUBLICATIONS: Research results, books, and specialty publications.

SWISS FEDERAL INSTITUTE OF TECHNOLOGY, LABORATORY FOR BUILDING MATERIALS SCIENCE, LAUSANNE

Chemin de Bellerive 32
CH-1007 Lausanne

CONTACT: Folker H. Wittmann, Director

MISSION: To provide education and advance knowledge of materials performance and characterization.

PRIMARY WORK: Performs research on concrete, fracture mechanics, material durability, prediction of behavior, and testing.

SOURCE OF FINANCES: Government, university, and industry.

PUBLICATIONS: Research reports, professional journal articles, and specialty publications.

FEDERAL HOUSING RESEARCH COMMISSION (CRL)

Weltpoststrasse 4
CH-3000 Bern 15

CONTACT: Peter Gurtner, Head

MISSION: To promote and fund research in the housing and building field and to disseminate findings to the public.

PRIMARY WORK: Performs work in modernization and rehabilitation, housing, economics, market analysis, and forecasting.

SOURCE OF FINANCES: Government.

PUBLICATIONS: Series, research reports, and specialty publications.

SWISS FEDERAL LABORATORIES FOR MATERIALS TESTING AND RESEARCH (EMPA)

Ueberlandstrasse 129
Postfach, CH-8600 Duebendorf
Zurich

CONTACT: H. M. Fischer, Public Relations

MISSION: Central testing and research organization that develops guidelines, standards, and regulations for building materials and manufactured products.

PRIMARY WORK: Investigations into the chemical and physical properties of buildings and manufactured materials, fuels, manufactured products and structures; testing of specimens and structural elements; nondestructive investigations; acceptance tests in accordance with national and international standards; and research work in the field of new materials.

SOURCE OF FINANCES: Fees for services and government.

PUBLICATIONS: Research reports and scientific publications in the specific literature.

SWISS FEDERAL INSTITUTE OF TECHNOLOGY
   APPLIED STATICS AND STEEL STRUCTURES (BS)
   INSTITUTE OF STRUCTURAL ENGINEERING (IBK)
   INSTITUTE FOR BUILDING MATERIALS, MATERIAL CHEMISTRY, AND CORROSION (IBWK)
   NSTITUTE OF FOUNDATION ENGINEERING AND SOIL MECHANICS (IGB)
   INSTITUTE FOR BUILDING RESEARCH (IBR)

ETH-Honggerberg
Gebaude HIL
CH-8093 Zurich

CONTACT: P. Dubas, Director, BS
           H. Bachmann, Head, IBK
           H. Bohni, Director, IBWK
           H. J. Lang, Director, IGB
           B. Hubert, Director, IBR

MISSION: To provide education and advanced building technology.

PRIMARY WORK: Performs research in structural engineering, steel and wood structures (BS); structural engineering, reinforced and prestressed concrete, earthquake engineering and dynamics, informatics (IBK); building materials, concrete, metals, reinforced plastics, corrosion (IBWK); civil engineering, foundations, soil mechanics, failure analysis, soil structures, clay mineralogy (IGB); housing, human sciences, renovation, quality control, energy requirements and conservation, economics (IBR).

SOURCE OF FINANCES: Government, fees for services, and industry.

PUBLICATIONS: Research reports, articles in professional journals, proceedings, technical recommendations, and specialty publications.

SWISS BUILDING DOCUMENTATION CENTER

CH-4249 Blauen

CONTACT: C. Weisser, Managing Director

MISSION: Central information center for the building industry.

PRIMARY WORK: Performs work in information documentation systems, data bases on products and services, and information services.

SOURCE OF FINANCES: Subscription fees and fees for services.

PUBLICATIONS: Monthly bulletin, catalogues, product information, files, and specialty publications.

SWISS INSTITUTE OF STEEL CONSTRUCTION (SZS)

Seefeldstrasse 25
CH-8034 Zurich

CONTACT: U. Wyss, Director

MISSION: To promote use of steel for construction.

PRIMARY WORK: Performs work in civil engineering, materials, standardization, and information systems; provides advice and training.

SOURCE OF FINANCES: Member firms.

PUBLICATIONS: Bimonthly journal and list of publications.

TECHNICAL RESEARCH AND ADVISORY INSTITUTE OF THE SWISS CEMENT INDUSTRY

Lindenstrasse 10
CH-5103 Wildegg

CONTACT: W. Wilk, Director

MISSION: Undertakes research in all fields (other than road construction) of hydraulic binders, mortar, aggregates, and concrete.

PRIMARY WORK: Advising, consulting, research, and testing in the field of all building materials containing hydraulic binders.

SOURCES OF FINANCES: Contributions from the Swiss Cement Industry.

PUBLICATIONS: *Cement Bulletin*, publications in professional journals, and specialty publications.

TANZANIA

NATIONAL CONSTRUCTION COUNCIL

P.O. Box 40465
Dar es Salaam

MISSION: To promote development of good building practices for the construction industry in order to maintain solid national growth.

PRIMARY WORK: Performs work in building materials, construction techniques and management, building design, and consultation and arbitration.

SOURCE OF FINANCES: Government.

PUBLICATIONS: Seminar reports and specialty publications.

NATIONAL HOUSING AND BUILDING RESEARCH UNIT

P.O. Box 1964
Dar es Salaam

CONTACT: A. L. Mtui, Director

MISSION: To improve building technology for affordable housing using indigenous materials.

PRIMARY WORK: Performs work in materials, soils, testing, low-cost housing, human requirements, economics, sociocultural conditions, and information systems.

SOURCE OF FINANCES: Government and foreign grants.

PUBLICATIONS: Research results and list of publications.

TURKEY

ISTANBUL TECHNICAL UNIVERSITY, MATERIALS AND STRUCTURES LABORATORIES

Insaat Fakultesi Ayazaga
Maslak, 80626
Istanbul

CONTACT: Mehmet Uyan, Director

MISSION: To provide education and improve technology for building materials and components.

PRIMARY WORK: Performs research on mechanical properties and service life predictions of concrete, ferrocement, gypsum, roofing products, and testing.

SOURCE OF FINANCES: University, grants, and fees for services.

PUBLICATIONS: Technical reports, abstracts, and Ph.D. theses.

SCIENTIFIC AND TECHNICAL RESEARCH COUNCIL OF TURKEY, BUILDING RESEARCH INSTITUTE

Bilir Sok 17
Kavaklidere
Ankar

CONTACT: Mustafa Pultar, Director

MISSION: National research laboratory conducting building research and providing advisory services.

PRIMARY WORK: Performs research in civil and seismic engineering, materials, energy conservation, building physics, fire protection, economics, construction management, quality control, information systems, and testing.

SOURCE OF FINANCES: Government and sale of publications.

PUBLICATIONS: Journals, research results, symposia series, and specialty publications.

USSR

ALL-UNION RESEARCH INSTITUTE FOR INFORMATION ON CONSTRUCTION AND ARCHITECTURE

Gosstroy
12 Marx Avenue
103828 Moscow

CONTACT: I. E. Evgenyev, Director

MISSION: To process and prepare building technology information for dissemination.

PRIMARY WORK: Performs work in information systems, building catalogue, and information analysis.

SOURCE OF FINANCES: Government and fees for services.

PUBLICATIONS: Semiweekly abstract journals, bulletins, and reviews.

CENTRAL RESEARCH, DESIGN AND EXPERIMENTAL INSTITUTE FOR INDUSTRIAL BUILDINGS AND STRUCTURES AND FOR ORGANIZATION, MECHANIZATION, AND TECHNICAL ASSISTANCE IN CONSTRUCTION

Gosstroy
12 Marx Avenue
103828 Moscow

CONTACT: Yu N. Khromets, Director, Industrial Buildings and Structures
E. A. Dolginin, Director, Organization, Mechanization, and Technical Assistance in Construction

MISSION: Central source for testing materials and components, for developing technology for industrial buildings and structures, and for organizing, mechanizing, and automating construction.

PRIMARY WORK: Performs work in testing, architecture, planning, roofing, standardization, CAD, fire protection, HVAC, quality control, civil engineering, construction organization, concrete, and safety engineering.

SOURCE OF FINANCES: Government.

PUBLICATIONS: Reports on regulations and specialty publications.

RESEARCH INSTITUTE FOR CONCRETE AND REINFORCED CONCRETE

Gosstroy
12 Marx Avenue
103828 Moscow

CONTACT: K. V. Mikhailov, Director

MISSION: To improve the methods of analysis of concrete and reinforced concrete structures.

PRIMARY WORK: Performs research on all types of concretes, reinforced concrete, construction techniques, protection, corrosion, durability, economics, and fire resistance.

SOURCE OF FINANCES: Government and fees for services.

PUBLICATIONS: Monthly journal.

RESEARCH INSTITUTE FOR BUILDING PHYSICS

Gosstroy
12 Marx Avenue
103828 Moscow

CONTACT: V. A. Drozdov, Director

MISSION: To improve building physics technologies.

PRIMARY WORK: Performs research in thermal engineering, lighting, acoustics, climatology, moisture control, and economics.

SOURCE OF FINANCES: Government and fees for services.

PUBLICATIONS: Monographs and specialty publications.

CENTRAL RESEARCH INSTITUTE FOR BUILDING STRUCTURES

Gosstroy
12 Marx Avenue
103828 Moscow

CONTACT: A. F. Smirnov, Director

MISSION: To improve building mechanics knowledge.

PRIMARY WORK: Performs research on civil and seismic engineering, computing, test methods, and fire protection.

SOURCE OF FINANCES: Government and fees for services.

PUBLICATION: Bimonthly journal.

UNITED KINGDOM

ASLIB

3 Belgrave Square
London SW1X 8PL

CONTACT: Dennis A. Lewis, Director

MISSION: To process and disseminate building technology information.

PRIMARY WORK: Performs work in information handling and retrieval systems, training, and referral services.

SOURCE OF FINANCES: Member subscriptions, grants, and sales of publications and services.

PUBLICATIONS: Monthly and quarterly journals, index, handbooks, directories, monographs, research reports, and bibliographies.

BRIGHTON POLYTECHNIC, BUILDING DEPARTMENT AND CIVIL ENGINEERING DEPARTMENT

Mithras House,
Lewes Road
Moulsecoomb, Brighton

CONTACT: P. J. Rutland, Head, Building Department
Barry W. Staynes, Director, Civil Engineering Department

MISSION: To provide education and research to improve building technology.

PRIMARY WORK: Performs research in environmental sciences, building services, materials, material sciences, concrete, timber, economics, project control, construction management, civil engineering, geotechnical engineering, traffic engineering, CAD, testing, and information systems.

SOURCE OF FINANCES: Government.

PUBLICATIONS: Research reports and papers.

BRISTOL POLYTECHNIC, DEPARTMENT OF CONSTRUCTION AND ENVIRONMENTAL HEALTH

Ashley Down Road
Bristol BS7 9BU

CONTACT: M. M. Cusack, Director

MISSION: To educate and advance construction technologies.

PRIMARY WORK: Performs work in environmental health, management, and organization.

SOURCE OF FINANCES: Government.

PUBLICATIONS: Annual research report.

THE BRITISH BOARD OF AGREMENT (BBA)

P.O. Box 195
Bucknalls Lane
Garston, Watford, Herts WD2 7NG

CONTACT: T. P. R. Lant, Director

MISSION: To provide for quality control of the construction industry through independent assessment of materials, products, components, and processes.

PRIMARY WORK: Performs work in evaluating materials, timber, insulations, structures, cladding, floors, windows, condensation, and thermal performance; testing; and issues certificates.

SOURCE OF FINANCES: Fees for services and government grants.

PUBLICATIONS: Agrement certificates, assessment and test methods, information sheets, and newsletter.

BRITISH CERAMIC RESEARCH ASSOCIATION, LTD. (CERAM RESEARCH)

Queens Road
Penkhull, Stoke-on-Trent
Staffs ST4 7LQ

CONTACT: D. W. F. James, Chief Executive

MISSION: Research and development, consultancy, testing services, and information on ceramics and inorganic materials.

PRIMARY WORK: Provides contract research, consultancy, and testing into ceramics and inorganic materials for all segments of the ceramic industry including: ceramic tableware; sanitaryware; roof, floor, and wall tiles; masonry products; refractories; and advanced ceramics.

SOURCE OF FINANCES: Member programs, fees for contract research, consultancy, testing, and information services.

PUBLICATIONS: Quarterly journal, quarterly newsletter, technical standards, research reports, catalogues, and monthly abstracts.

BUILDING CENTER

26 Store Street
London WC1E 7BT

CONTACT: John F. George, Chief Executive

MISSION: To provide information to specifiers and users of building materials and services.

PRIMARY WORK: Performs services through permanent display of products, advisory service, information systems, retrieval services, and dissemination methods.

SOURCE OF FINANCES: Fees for services.

PUBLICATIONS: Facts sheets.

BUILDING RESEARCH ESTABLISHMENT (BRE)

Garston, Watford WD2 7JR

CONTACT: Rex Watson, Director

MISSION: National research laboratory that develops advanced building technology for building and construction and for the prevention and control of fires.

PRIMARY WORK: Performs research in civil engineering, geotechnical engineering, construction, fire protection, building regulations, energy conservation, thermal engineering, acoustics, lighting, building services, materials, concrete, and information systems.

SOURCE OF FINANCES: Government with some private industry sponsorship.

PUBLICATIONS: Monthly journals, research reports, catalogues, annual report, monthly digests, defect sheets, and information papers.

BUILDING SERVICES RESEARCH AND INFORMATION ASSOCIATION

Old Bracknell Lane West
Bracknell
Berks RG12 4AH

CONTACT: D. P. Gregory, Director

MISSION: To provide members with information and research services on improving building services technology.

PRIMARY WORK: Performs research on energy conservation and use, air movement, lighting, ventilation, building communication systems; maintains information and documentation services; and conducts specialized consultations and testing.

SOURCE OF FINANCES: Member subscriptions and fees for services.

PUBLICATIONS: Research reports, technical notes, bimonthly abstracts, bibliographies, and guides.

CEMENT AND CONCRETE ASSOCIATION (C&CA)

Wexham Springs
Slough SL3 6PL

CONTACT: R. E. Rowe, Director-General

MISSION: To serve cement industry members with advanced building technology, research and advisory services.

PRIMARY WORK: Performs work in cement, concrete, masonry, testing, advisory services, and training.

SOURCE OF FINANCES: Cement industry members.

PUBLICATIONS: Quarterly journal and magazine, technical reports, and advisory publications.

CHARTERED INSTITUTION OF BUILDING SERVICES

Delta House
222 Balham High Road
Balham
London SW12 9BS

CONTACT: A. V. Ramsey, Secretary

MISSION: To promote the improvement of building services engineering techniques and practices.

PRIMARY WORK: Professional qualification and registration, accreditation of courses, examinations, technical publications, meetings, seminars in fields of heating, ventilation, air conditioning, lighting, energy use, electrical services, and mechanical services.

SOURCE OF FINANCES: Membership and sale of publications.

PUBLICATIONS: Technical handbooks, data books, monthly and quarterly journals, research reports, and symposium proceedings.

CONSTRUCTION INDUSTRY RESEARCH AND INFORMATION ASSOCIATION (CIRIA)

6 Storey's Gate
Westminister
London SW1P 3AU

CONTACT: P. L. Bransby, Director

MISSION: To identify research needs, manage research, and disseminate results for the construction industry.

PRIMARY WORK: Performs work in structural design, civil engineering, construction, foundation engineering, and underground construction, offshore construction.

SOURCE OF FINANCES: Subscriptions and fees for research.

PUBLICATIONS: Research reports, technical reports, bimonthly newsletter, annual report, and bibliographies.

COVENTRY POLYTECHNIC, DEPARTMENT OF CIVIL ENGINEERING AND BUILDING

Priory Street
Coventry, West Midlands CV1 5FB

CONTACT: H. E. Walker, Head

MISSION: To educate and advance building technology.

PRIMARY WORK: Performs research in soil mechanics, civil engineering, reinforced brickwork, energy conservation, air flows, economics, acoustics, building performance, and construction management.

SOURCE OF FINANCES: Government, fees for services, and Science and Engineering Research Council.

PUBLICATIONS: Triennial research report, research results, and specialty publications.

HERIOT-WATT UNIVERSITY, DEPARTMENT OF BUILDING

Chambers Street
Edinburgh, EH1 1HY

CONTACT: Victor Brownlie Torrance, Head

MISSION: To provide academic education and advance the state of building technology.

PRIMARY WORK: Performs research on construction management, acoustics, energy conservation and use, CAD, air flows, building services, economics, architecture, and environmental studies.

SOURCE OF FINANCES: University, grants, government, and fees for services.

PUBLICATIONS: Research reports, monographs, and specialty publications.

IMPERIAL COLLEGE OF SCIENCE AND TECHNOLOGY, DEPARTMENT OF CIVIL ENGINEERING

London SW7 2BU

CONTACT: Patrick Dowling, Head

MISSION: To teach and and advance the state of civil engineering and building technology.

PRIMARY WORK: Performs research on steel and concrete structures (onshore and offshore), soil mechanics, earthquake engineering, hydraulics, and water technology.

SOURCE OF FINANCES: Government and industry.

PUBLICATIONS: Research reports.

LAING TECHNOLOGY GROUP, LTD.

Page Street
Mill Hill
London NW7 2ER

CONTACT: P. D. Westwood, Director

MISSION: To provide improved technology and support services to Laing Design and to the building industry.

PRIMARY WORK: Performs work in civil engineering, energy use, construction, quality assurance, product testing, training, geotechnical engineering, and soil mechanics.

SOURCE OF FINANCES: Corporation and fees for services from industry.

PUBLICATIONS: Research reports for clients.

LEEDS POLYTECHNIC, SCHOOL OF CONSTRUCTION STUDIES

Brunswick Terrace
Leeds LS2 8BU

CONTACT: D. W. H. Skinner

MISSION: Education and research to improve building technology.

PRIMARY WORK: Performs work in civil engineering, environmental health, construction, thermal engineering, test methods, and building renovation.

SOURCE OF FINANCES: Government.

PUBLICATIONS: Research reports, brochures, text books, and specialty publications.

LEICESTER POLYTECHNIC, SCHOOL OF ARCHITECTURE

P.O. Box 143
Leicester LE1 9BH

CONTACT: Theo Matoff, Director

MISSION: To educate and perform research to advance building practices.

PRIMARY WORK: Environmental studies and CAD.

SOURCE OF FINANCES: University.

PUBLICATIONS: Semiannual journal and research reports.

MATERIALS AND COMPONENTS DEVELOPMENT AND TESTING ASSOCIATION

Paisley College Of Technology
47 High Street
Paisley PA1 2BE

CONTACT: E. Downey, Director

MISSION: Education and research for construction practices.

PRIMARY WORK: Performs work on materials, structures, product testing and certification, and acoustics.

SOURCE OF FINANCES: Fees for services.

PUBLICATIONS: Research reports.

NEW UNIVERSITY OF ULSTER

Colerine
N. Ireland BT52 1SA

CONTACT: J. T. McMullan and R. Morgan, Directors

MISSION: To develop technology to improve the quality of building practices.

PRIMARY WORK: Performs research on building physics, heat pumps, controls, heat recovery, and energy use.

SOURCE OF FINANCES: University, government, and grants.

PUBLICATIONS: Specialty publications.

OVE ARUP & PARTNERS

13 Fitzroy Street
London W1P 6BQ

CONTACT: Povl Ahm, Chairman

MISSION: To provide consultation in engineering and analysis of building structures.

PRIMARY WORK: Performs work in civil engineering, engineering design, acoustics, fire engineering, mechanical engineering, and economics.

SOURCE OF FINANCES: Fees for services.

PUBLICATION: Quarterly journal.

OXFORD POLYTECHNIC, DEPARTMENT OF CIVIL ENGINEERING BUILDING AND CARTOGRAPHY

Gypsy Lane
Headington, Oxford OX3 0BP

CONTACT: R. W. Morris, Head

MISSION: To undertake research to advance building and civil engineering practices.

PRIMARY WORK: Provides research on materials, concrete, soil mechanics, structures, hydraulics, hydrology, and testing.

SOURCE OF FINANCES: Fees for research, consultancy, and grants.

PUBLICATIONS: Professional journals.

PLYMOUTH POLYTECHNIC, DEPARTMENT OF CIVIL ENGINEERING

Place Court
Palace Street
Plymouth PL1 2DE

CONTACT: R. J. Cope, Head

MISSION: Education and research in civil and structural engineering, including building technology.

PRIMARY WORK: Performs research on civil engineering, materials, concrete, reinforced brickwork, soil mechanics, hydraulics, highway maintenance, coastal engineering, and underwater technology.

SOURCE OF FINANCES: Government and fees for services.

PUBLICATIONS: Journal and research reports.

POLYTECHNIC OF THE SOUTH BANK, FACULTY OF THE BUILT ENVIRONMENT

Wandsworth Road
London SW8

CONTACT: Patricia Roberts, Dean

MISSION: Education to advance building practices.

PRIMARY WORK: Performs research on civil engineering, town planning, economics, organization and management, and construction.

SOURCE OF FINANCES: University and fees for research.

PUBLICATIONS: Technical standards, research reports, and bibliographies.

PORTSMOUTH POLYTECHNIC, DEPARTMENT OF CIVIL ENGINEERING

Burnaby Building
Burnaby Road
Portsmouth PO1 3QL

CONTACT: P. A. Stead, Head of Department

MISSION: To advance the development of new techniques and systems.

PRIMARY WORK: Performs research on materials engineering, structural engineering, geotechnical engineering, construction, transportation, hydraulics, CAD, transportation systems, and waste water treatment.

SOURCE OF FINANCES: University, government, industry, and fees for research.

PUBLICATIONS: Research reports and text books.

REDLAND TECHNOLOGY LIMITED

Graylands, Langhurstwood Road
Horsham
West Sussex RH12 4QG

CONTACT: S. E. Horsley, Managing Director

MISSION: To develop new technology and products.

PRIMARY WORK: Provides work on materials, ceramics, concrete, polymers, economics, energy, road surfaces, wind tunnel, and environmental testings.

SOURCE OF FINANCES: Member companies.

PUBLICATIONS: Specialty publications and technical reports.

ROYAL INSTITUTE OF BRITISH ARCHITECTS (RIBA)

66 Portland Place
London W1N 4AD

CONTACT: Donald Brooks, Acting Secretary

MISSION: To advance civil architecture and to promote and facilitate the acquirement of knowledge of the various arts and sciences connected therewith.

PRIMARY WORK: Performs research into specific aspects of design and construction, training, forums, and information systems.

SOURCE OF FINANCES: Members and services to members.

PUBLICATIONS: Monthly and quarterly journals, indexes, books, and specialty publications.

THE STEEL CONSTRUCTION INSTITUTE

Silwood Park
Ascot
Berkshire SL5 7QN

CONTACT: C. J. Billington

MISSION: To develop and promote the proper and effective use of steel as a construction material.

PRIMARY WORK: Develops design recommendations for steel construction, fire protection, etc.; provides advisory services, information services, quality assurance, computer aids, technical education for universities and practitioners, and technical publications; and performs research into new techniques and uses for steel in construction.

SOURCE OF FINANCES: Membership subscriptions, sales of publications, and research contracts.

PUBLICATIONS: Design guidelines, handbooks, monographs, and research results.

TIMBER RESEARCH AND DEVELOPMENT ASSOCIATION (TRADA)

Stocking Lane
Hughenden Valley
High Wycombe
Buckinghamshire HP14 4ND

CONTACT: J. G. Sunley, Director

MISSION: To perform research to advance timber practices.

PRIMARY WORK: Provides work in construction, timber engineering, fire engineering, quality assurance, and testing.

SOURCE OF FINANCES: Members, levy from imported timber trade, and fees for services.

PUBLICATIONS: Recommendations, information sheets, books, and bibliographies.

TRANSPORT AND ROAD RESEARCH LABORATORY (TRRL)

Crowthorne
Berkshire RG11 6AU

CONTACT: R. Bridle, Director

MISSION: To advance technology to formulate, develop, and implement government road and transport policies.

PRIMARY WORK: Performs research on civil engineering, construction engineering, highway engineering, urban and regional planning, and economics.

SOURCE OF FINANCES: Government.

PUBLICATIONS: Research reports, papers in scientific journals, and specialty publications.

UNIVERSITY COLLEGE LONDON, BARTLETT SCHOOL OF ARCHITECTURE AND PLANNING

Wates House
22 Gordon Street
London WC1H 0QB

CONTACT: John Andrews, Director

MISSION: Education to advance the state of knowledge of construction, architecture, planning, and economics.

PRIMARY WORK: Performs work in architecture, construction, housing, economics, urbanization, and low-cost buildings.

SOURCE OF FINANCES: Government, university, and fees for research.

PUBLICATIONS: Research reports, bibliographies, theses, and papers in journals.

UNIVERSITY COLLEGE LONDON, DEPARTMENT OF CIVIL ENGINEERING

Gower Street
London WC1E 6BT

CONTACT: K. O. Kemp, Head

MISSION: Education to advance civil engineering technologies.

PRIMARY WORK: Performs research on geotechnical engineering, structural mechanics, offshore structural engineering, concrete and concrete structures, and fluid mechanics.

SOURCE OF FINANCES: University and government.

PUBLICATIONS: Research and project reports.

BATH UNIVERSITY, SCHOOL OF ARCHITECTURE AND BUILDING ENGINEERING

Clavertyon Down,
Bath BA2 7AY

CONTACT: Michael Brawne, Director

MISSION: Education to advance the state of technology in building sciences.

PRIMARY WORK: Performs research on architecture, environmental engineering, construction, acoustics, lighting, structures, soil mechanics, and civil engineering.

SOURCE OF FINANCES: University and grants.

PUBLICATIONS: Research reports and papers in journals.

BIRMINGHAM UNIVERSITY, CIVIL ENGINEERING DEPARTMENT

P.O. Box 363
Birmingham B15 2TT

CONTACT: B. P. Hughes, Acting Head

MISSION: Education to advance the state of technology in civil engineering.

PRIMARY WORK: Performs research on structures, materials, concrete, fluid mechanics, hydrology, public health, soil mechanics, and foundation engineering.

SOURCE OF FINANCES: University research grants and fees for services.

PUBLICATIONS: Research reports, annual report, and biannual research reports.

EDINBURGH UNIVERSITY, DEPARTMENT OF CIVIL ENGINEERING AND BUILDING SCIENCE

The King's Buildings
Mayfield Road
Edinburgh EH9 3JL

CONTACT: J. Morgan, Head

MISSION: Education to advance civil engineering practices.

PRIMARY WORK: Performs research in structural masonry, ceramics, nondestructive testing, timber structures, geotechnical engineering, concrete, and wind loads.

SOURCE OF FINANCES: University and grants.

PUBLICATIONS: Research reports, proceedings, and papers in journals.

LIVERPOOL UNIVERSITY, SCHOOL OF ARCHITECTURE AND BUILDING ENGINEERING

P.O. Box 147
Liverpool L69 3BX

CONTACT: D. W. Cheetham, Director of Studies in Building Engineering

MISSION: Education to advance technology for better building practices.

PRIMARY WORK: Performs research on building construction engineering, acoustics, energy consumption, building services, lighting, materials, concrete, and productivity.

SOURCE OF FINANCES: University and grants.

PUBLICATIONS: Papers in journals.

MANCHESTER UNIVERSITY,
INSTITUTE OF SCIENCE AND TECHNOLOGY,
DEPARTMENT OF BUILDING

P.O. Box 88
Manchester M60 1QD

CONTACT: R. Pilcher and P. J. Burberry, Co-Directors

MISSION: Education to advance building services and construction engineering technologies.

PRIMARY WORK: Performs research on materials, construction, nondestructive testing, building design, econometrics, and housing.

SOURCE OF FINANCES: University, grants, and fees for services.

PUBLICATIONS: Research reports, bibliographies, and papers in journals.

NEWCASTLE UPON TYNE,
SCHOOL OF ARCHITECTURE,
BUILDING SCIENCE SECTION

The University
Claremont Road
Newcastle upon Tyne NE1 7RU

CONTACT: A. C. Hardy, Professor of Building Science

MISSION: Education to advance building technology.

PRIMARY WORK: Undertakes research into building physical performance, energy conservation, environmental standards, natural and artificial lighting, building acoustics, and building performance prediction techniques by computer simulation.

SOURCE OF FINANCES: External funding from research councils and statutory bodies.

PUBLICATIONS: Research reports available from British Scientific Library.

NEWCASTLE UPON TYNE UNIVERSITY, DEPARTMENT OF CIVIL ENGINEERING

The University
Claremont Road
Newcastle upon Tyne NE1 7RU

CONTACT: M. B. Pescod, Head, Civil Engineering

MISSION: Education to advance the state of building technology.

PRIMARY WORK: Performs research on structural, environmental, and hydraulic engineering.

SOURCES OF FINANCE: University and grants.

PUBLICATIONS: Research reports and papers in journals.

SALFORD UNIVERSITY, DEPARTMENT OF CIVIL ENGINEERING

Salford M5 4WT

CONTACT: E. R. Bryan, Chairman

MISSION: Education to advance the knowledge of the built environment.

PRIMARY WORK: Performs research and consultancy on civil engineering, quantity surveying, building surveying, environmental sciences, housing materials, construction, CAD, testing, and management.

SOURCE OF FINANCES: University and grants.

PUBLICATIONS: Research reports and papers in journals.

SHEFFIELD UNIVERSITY, DEPARTMENT OF CIVIL AND STRUCTURAL ENGINEERING

Mappin Street
Sheffield S1 3JD

CONTACT: T. H. Hanna, Head

MISSION: Education to advance building practices.

PRIMARY WORK: Performs research on civil, geotechnical, transport and traffic engineering, and structural steel and concrete.

SOURCE OF FINANCES: University and government.

PUBLICATIONS: Papers in journals and annual report.

**STRATHCLYDE UNIVERSITY,
DEPARTMENT OF ARCHITECTURE AND BUILDING SCIENCES**

131 Rottnrow
Glasgow G4 0NG

CONTACT: P. P. Yaneske

MISSION: Education and advanced training to apply building technology to the efficient design of buildings.

PRIMARY WORK: Research on building performance, CAD, construction, energy conservation, housing rehabilitation, information management and CAD systems.

SOURCE OF FINANCES: University, government, industry professions, and grants.

PUBLICATIONS: Biannual bulletin, research reports, occasional paper series, and specialty publications.

**UNIVERSITY OF WALES,
INSTITUTE OF SCIENCE AND TECHNOLOGY,
WELSH SCHOOL OF ARCHITECTURE, RESEARCH, AND DEVELOPMENT**

20-22 North Road
Cardiff, CF1 3DY

CONTACT: J. E. Eynon, Head

MISSION: Education to advance environmental sciences, user attitudes, and design techniques.

PRIMARY WORK: Performs research on acoustics, CAD, lighting, space use, and energy conservation.

SOURCE OF FINANCES: University and grants.

PUBLICATIONS: Research reports and papers in journals.

PROPERTY SERVICES AGENCY (PSA)

Whitgift Center
Wellesley Road
Croydon CR9 3LY

CONTACT: Charles Rogers, Chief Librarian

MISSION: To provide, manage, maintain, and furnish information about building design and construction, and to design and supervise government construction.

PRIMARY WORK: Performs work in real estate management, design, construction, information documentation and retrieval systems, and innovations to improve productivity.

SOURCE OF FINANCES: Government.

PUBLICATIONS: Quarterly journal, research reports, and list of publications.

VENEZUELA

NATIONAL INSTITUTE OF HOUSING

Edificio Cruz Verde
Esquina Cruz Verde
Caracas 1010

CONTACT: Tito Fernando Herrera Armas, President

MISSION: To encourage development of technology to improve low-cost housing.

PRIMARY WORK: Provides services in the management of projects, housing, land, rehabilitation, economics, and quality control.

SOURCE OF FINANCES: Government, grants, and banks.

PUBLICATIONS: Program reports, policies and recommendations, and specialty publications.

**CENTRAL UNIVERSITY OF VENEZUELA,
INSTITUTE OF MATERIALS AND STRUCTURAL MODELS (IMME)**

Apdo 50.361
Caracas 1050A

CONTACT: Maria L. Diaz de Smitter, Director

MISSION: Education to improve technologies for the construction industry.

PRIMARY WORK: Performs research on low-cost buildings, construction, indigenous materials, earthquake engineering, precast concrete, and information systems.

SOURCE OF FINANCES: University and fees for services.

PUBLICATIONS: Biannual journal, research reports, and specialty publications.

YUGOSLAVIA

HIGHWAY INSTITUTE

Kumodrasks 257
11000 Belgrade

MISSION: To provide research for highway and traffic engineering.

PRIMARY WORK: Performs research on geotechnical engineering, concrete, asphalt, soil mechanics, foundation engineering, quality control, transport design, and construction.

SOURCE OF FINANCES: Fees for services.

PUBLICATIONS: Semiannual journal, research reports, standards, and specialty publications.

INSTITUTE FOR METAL STRUCTURES

Memcingerjeva 7
6100 Ljubljana

CONTACT: Ciril Sivic, Director

MISSION: To develop advances in metals technologies.

PRIMARY WORK: Performs research on structural engineering, welding technology, nondestructive testing, materials, and information systems.

SOURCE OF FINANCES: Fees for services and grants.

PUBLICATIONS: Annual information bulletins and specialty publications.

INSTITUTE FOR RESEARCH AND TESTING IN MATERIALS AND STRUCTURES

Dimiceva 12
Ljubljana

CONTACT: Joze Vizintin, Director

MISSION: To advance building technology.

PRIMARY WORK: Performs research on building materials, structures, properties, corrosion, thermal engineering, acoustics, fire protection, civil engineering, cement and concrete, bridges, transport, and testing.

SOURCE OF FINANCES: Government and fees for services.

PUBLICATIONS: Research reports, monthly information bulletin, annual reports, bibliographies, and catalogues.

INSTITUTE OF SR SERBIA FOR TESTING MATERIALS

Bulevar vojvode Misica br. 43
11000 Belgrade

CONTACT: Milos Banic, Director

MISSION: To perform testings of building components and products.

PRIMARY WORK: Provides services in materials and construction, innovative systems, precast construction, and quality control.

SOURCE OF FINANCES: Fees for services, grants, and government.

PUBLICATIONS: Periodic bulletins and specialty publications.

YUGOSLAV BUILDING CENTER

84 Bulevar revolucije
11000 Belgrade

CONTACT: Dragomir Maricic, General Manager

MISSION: To improve and develop building construction and materials through maintaining a central repository of information.

PRIMARY WORK: Performs work on information collection, storage, and documentation systems; training; educating; and exhibition.

SOURCE OF FINANCES: Grants, fees for service, and sale of documents.

PUBLICATIONS: Research reports, monthly journal, technical recommendations, standards, and regulations.

Index of Organizations

A

Aachen Technical University, Institute for Building Research (West Germany), 114
Academy of Building of the German Democratic Republic, 127
ACEC Research & Management Foundation, 3
Acid Deposition Research Program, 58–59
Acid Rain Research Program, 59
Acoustics Laboratory (Norway), 163
Agrement Board of South Africa, 176
Agroman Empresa Constructor S.A., Division of Research and Methods (Spain), 178
Air Energy Engineering Research Laboratory, 59
Alaska, University of, 40–41
All-Union Research Institute for Information on Construction and Architecture (USSR), 192
American Concrete Institute, 4
American Council of Independent Laboratories, 4
American Gas Association, 23
American Institute of Architects, 4
American Institute of Steel Construction, 5
American Iron & Steel Institute, 5
American Plywood Association, 6
American Society of Civil Engineers, 6
American Society of Heating, Refrigerating, and Air-Conditioning Engineers, 7
American Society of Plumbing Engineers Research Foundation, 7
Architectural Institute of Korea, 151
Argentina Portland Cement Institute, 77
Aristotle Thessaloniki University, School of Engineering, Department of Civil Engineering, Division of Structural Engineering (Greece), 128
Armstrong World Industries, Inc., 14
ASLIB (United Kingdom), 194
Asphalt Institute Research Center, The, 7
Association of Finnish Civil Engineers, 120–121
ASTM, 8
Auckland University, School of Architecture (New Zealand), 162
Australian Institute of Steel Construction, 77
Australian Road Research Board, 77
Austrian Cement Research Institute, 81
Austrian Institute for Building Research, 81
Austrian Plastics Institute, 82–83
Austrian Wood Research Institute, 82
Autodesk, Inc., 14
Auto-Trol Corporation, 14

B

Bari University, Institute of Engines and Energetics, Laboratory of Thermoenergetics (Italy), 142
Bari University, Institute of Technical Physics and Thermotechnical Plants (Italy), 142
BASF Corporate Research and Development Center, 15
Bath University, School of Architecture and Building Engineering (United Kingdom), 205–206
Battelle Memorial Institute, 23, 71
Battelle-Pacific Northwest Laboratories, 69
Belgian Building Research Center, 85
Belgium Center for Corrosion Study, 84–85
Berlin Technical University, Institute for Structural Design and Material Strength, Hermann Rietschel Institute for Heating and Air-Conditioning Engineering (West Germany), 116–117
Berlin Technical University, Institute of Soil Mechanics and Foundation Engineering (West Germany), 116

Birmingham University, Civil Engineering Department (United Kingdom), 206
Bolt, Beranek, and Newman Laboratories, Inc., 15
Bouwcentrum (Netherlands), 152
Braunschweig Technical University, Institute for Statics (West Germany), 117
Braunschweig Technical University, Institute of Building Materials, Reinforced Concrete and Fire Protection (West Germany), 117
Brick Development Research Institute (Australia/New Zealand), 78
Brick Institute of America, The, 8
Brick Laboratory of the Finnish Brick Industry Association, 121
Brighton Polytechnic, Building Department and Civil Engineering Department (United Kingdom), 195
Bristol Polytechnic, Department of Construction and Environmental Health (United Kingdom), 195
British Board of Agrement, 195–196
British Ceramic Research Association, Ltd., 196
Brookhaven National Laboratory, 60
Brussels Free University (Belgium), 89
Building and Cement Institute Eduardo Torroja (Spain), 179
Building and Civil Engineering Tests and Research Institute (Italy), 140
Building and Road Research Institute (Ghana), 128
Building Center (Italy), 139
Building Center of Japan, 143
Building Center (United Kingdom), 196
Building Contractors Society (Japan), 143–144
Building Design Institute for Typified Construction (Romania), 173
Building Division of the German Standards Institution (West Germany), 113–114
Building Industries Federation (South Africa), 176–177
Building Information Institute (Finland), 121
Building Institute for Testing Materials and Structures (Czechoslovakia), 100
Building Owners & Managers Association International, 9
Building Research Association of New Zealand, 159–160
Building Research Centre (Jordan), 149–150
Building Research Establishment (United Kingdom), 197
Building Research Foundation (Netherlands), 152
Building Research Institute (Czechoslovakia), 100
Building Research Institute (Iceland), 131
Building Research Institute (Japan), 144
Building Research Institute (Poland), 168–169
Building Research Station (Israel), 137
Building Research Station Lahore (Pakistan), 166–167
Building Services Research and Information Association (United Kingdom), 197
Building Standards Institution (Sweden), 180–181

Bureau of Mines, 58

C

Calgary University, Department of Civil Engineering (Canada), 96
California, University of, 30–31, 41–42
Calma Company, 16
Canada Mortgage and Housing Corporation, 91
Canadian Construction Management Institute, Canadian Construction Management Development Institute, 92
Canadian Institute of Steel Construction, 92
Cantabria University, School of Civil Engineering (Spain), 179
Canterbury University, Department of Civil Engineering (New Zealand), 162
Carnegie-Mellon University, 30
Carrier Corporation, 28
Caterpillar, Inc., 16
Catholic University of Leuven, Acoustics and Heat Conduction Laboratory (Belgium), 86–87
Catholic University of Leuven, Civil Engineering Department (Belgium), 87
Catholic University of Leuven, Department of Architecture (Belgium), 86–87
Cement and Concrete Association of Australia, 78
Cement and Concrete Association (United Kingdom), 197–198
Cement and Concrete Laboratory (Denmark), 102
Cement Research Institute of India, 132
Center for Building Studies (Canada), 92–93
Center for Building Technology, 61
Center for Fire Research, 63
Center for Research and Development in Masonry (Canada), 93
Central Building Research Institute (India), 132
Central Building Research Institute, Department of Cluj-Napoca (Romania), 174
Central Institute for Industrialization and Building Technology (Italy), 140
Central Research Design and Experimental Institute for Industrial Buildings and Structures and for Organization, Mechanization, and Technical Assistance in Construction (USSR), 193
Central Research and Design Institute for the Silicate Industry (Hungary), 129
Central Research Institute for Building Structures (USSR), 194
Chartered Institution of Building Services (United Kingdom), 198
China Academy of Building Research, 98
China Building Technology Development Center, 98
CIAD Association for Computer Application in Engineering (Netherlands), 153
Clay Brick Association of Canada, 93
Clemson University, 42
Coatings Research Institute (Belgium), 86

Cold Regions Research and Engineering
 Laboratory, 64
Colombian Institute of Cement Producers, 99–100
Colorado State University, 42–43
Colorado, University of, 31
Columbia University, 31
Commonwealth Scientific and Industrial Research
 Organization, Division of Building Research
 Organization (Australia), 79
Computing and Development Center for Data
 Processing in Civil Engineering (West
 Germany), 114
Concrete and Structural Research Institute
 (Denmark), 102–103
Concrete Association of Finland, 121–122
Concrete Institute of Australia, 77
Concrete Roads, Ltd. (Switzerland), 187
Construction Engineering Research Laboratory,
 64–65
Construction Industry Development Board
 (Singapore), 175
Construction Industry Research and Information
 Association (United Kingdom), 198
Construction Technology Laboratories, Inc., 9–10
Controlled Demolition International, 16–17
Corning Glass Works, 17
Coventry Polytechnic, Department of Civil
 Engineering and Building (United Kingdom),
 199
Czechoslovak Building Centre, 101

D

Danish Building Center, 103
Danish Building Research Institute, 103
Danish Corrosion Center, 103–104
Danish Fire Protection Association, 104
Danish Illumination Engineering Laboratory, 104
Danish Lime and Brick Laboratory, 104–105
Danish National Testing Board, 105
Dantest-National Institute for Testing and
 Verification (Denmark), 105
Darmstadt Technical University, Information
 Processing in Building and Civil Engineering
 Organizations (West Germany), 115
Darmstadt Technical University, Institute for
 Concrete Structures (West Germany), 115
Darmstadt Technical University, Institute of Steel
 Construction (West Germany), 115–116
Deere and Company, 18
Delft Geotechnics (Netherlands), 153–154
Delft Hydraulics Laboratory (Netherlands), 153
Delft University of Technology, Department of
 Civil Engineering (Netherlands), 154
Department of Work Science (Sweden), 181
Directorate of Building Research, Center for
 Research and Development on Human
 Settlements (Indonesia), 134–135
Directorate of Engineering and Services, 65

Documentation Information Office for Building,
 Architecture, and Town Planning (Romania),
 174–175
Dublin, University of, Trinity College (Ireland),
 136–137
Dutch Cement Industry Association, 154–155

E

Edison Electric Institute, 10
Edinburgh University, Department of Civil
 Engineering and Building Science (United
 Kingdom), 206
Eindhoven University of Technology, Department
 of Architecture, Building, and Planning,
 Materials and Structures Laboratory
 (Netherlands), 155
Electric Power Research Institute, 10
Electronics Industries Association, 25
Engineering and Services Center, 66
Environmental Monitoring Systems Laboratory,
 59
Essen University, Institute of Building Physics
 (West Germany), 117
Everett I. Brown Company, Architects and
 Engineers, 18–19

F

Factory Mutual Engineering and Research, 19
Federal Highway Research Institute (West
 Germany), 109
Federal Housing Research Commission
 (Switzerland), 188
Federal Waterway Engineering and Research
 Institute (West Germany), 109
Finland Technical Research Centre, Division for
 Building Technology and Community
 Development, 123
Florence University, Faculty of Architecture,
 Construction Department, Department of
 Processes and Methods in Building
 Production (Italy), 140–141
Florida A&M University, 32
Florida, University of, 43
Ford Motor Company Glass Division, 19
Forest Products Institute (Portugal), 171–172
Forest Products Laboratory, 67
Forest Products Research and Development
 Institute, Department of Science and
 Technology (Philippines), 167
Forest Research Institute, Forest Products
 Division (New Zealand), 160
Foundation for Architects' Research
 (Netherlands), 155
Fraunhofer Institute for Building Physics (West
 Germany), 111
Fraunhofer Institute for Wood Research, Wilhelm
 Klauditz Institut (West Germany), 108
French Central Hydraulics Laboratory, 125

French Precast Concrete Study and Research Center, 123

G

Gas Research Institute, 11, 23
General Organization for Housing, Building, and Planning Research (Egypt), 108
Georgia Institute of Technology, 32–33, 44
German Society of Masonry Construction, 110
Ghent State University, Faculty of Engineering, and Laboratory for Fuel Technology and Heat Transfer (Belgium), 87–88
Ghent State University, Magnel Laboratory for Reinforced Concrete (Belgium), 88
GTE Sylvania Products Corporation, 26
Guardian Industries, 20
Guayaquil Chamber of Construction (Ecuador), 107

H

Hannover, University of, Institute for Hydraulics and Coastal Engineering (West Germany), 118
Hannover University, Institute of Building Materials and Materials Testing (West Germany), 118
Harvard University, 33
Hazama-Gumi Technical Research Institute (Japan), 144
Health Effects Research Laboratory, 59–60
Helsinki University of Technology, Faculty of Surveying and Civil Engineering (Finland), 122
Heriot-Watt University, Department of Building (United Kingdom), 199
Highway Institute (Yugoslavia), 211
Honeywell Inc., 20–21
Housing Research and Development Unit (Kenya), 150
Hungarian Institute for Building Science, 130

I

Illinois, University of, 33–34, 44, 45
Illinois Institute of Technology, 34
IMO State Housing Corporation (Nigeria), 163
Imperial College of Science and Technology, Department of Civil Engineering (United Kingdom), 199
Indian Concrete Institute, 132–133
Indian Institute of Technology, Kanpur (India), 133
Indonesian Road Research Institute, 135
Information Center for Regional Planning and Construction (West Germany), 111–112
Information Center of Building (Hungary), 130
Institute for Building Materials, Technical Industrial Museum (Austria), 83
Institute for Building Mechanization and Rock Mining (Poland), 169
Institute for Building Research (West Germany), 112
Institute of Civil Engineering and Building Technology (West Germany), 113
Institute for Industrial Research and Standards, Construction Industry Division (Ireland), 136
Institute for Metal Structures (Yugoslavia), 211
Institute for Quality Control of Building (Hungary), 130–131
Institute for Reinforced Concrete (West Germany), 113
Institute for Research in Construction, National Research Council of Canada, 93–94
Institute for Research and Testing in Materials and Structures (Yugoslavia), 211
Institute of SR Serbia for Testing Materials (Yugoslavia), 212
Institute of Technology of the Construction of Cataluna (Spain), 178–179
Institution for Research and Material Testing (West Germany), 111
Intelligent Buildings Institute, 11–12
International Development Research Centre (Canada), 97
Interprofessional Technical Union of National Federations of Building and Public Works (France), 126–127
Iowa State University, 45–46
Istanbul Technical University, Materials and Structures Laboratories (Turkey), 191–192
Italian Technical and Economic Cement Association, 138
ITT Research Institute, 71–72

J

Jamaica Building Research Institute, 142–143
Johnson Controls, Inc., 21
Jordan, University of, Faculty of Engineering and Technology, Department of Civil Engineering, 150

K

Kajima Institute of Construction Technology (Japan), 144–145
Karlsruhe, University of, Chair of Soil Mechanics and Foundation Engineering, Institute of Soil Mechanics and Rock Mechanics (West Germany), 120
Karlsruhe University, Department of Steel and Aluminium Construction, Building Materials Testing Institute (West Germany), 120
Kawneer Company, Inc., 22
Kidde Automated Systems, 22
Koppers Company, Inc., 22
Korea Institute of Construction Technology, 151

L

Laboratory for Plastics Technology (Austria), 82
L.A. Falcao Bauer Testing Laboratory (Brazil), 90
Laing Technology Group, Ltd. (United Kingdom), 200
Lawrence Berkeley Laboratory, Center for Building Sciences, 67–68
Leeds Polytechnic, School of Construction Studies (United Kingdom), 200
Lehigh University, 46–48
Leicester Polytechnic, School of Architecture (United Kingdom), 200
Liege State University, Research Centers for Architecture and Town Planning, Civil Engineering, Building Physics Laboratory (Belgium), 88–89
Lennox Industries Inc., 23
Lighting Research Institute, 12
Liverpool University, School of Architecture and Building Engineering (United Kingdom), 206–207
L.M.I. B.V. Laboratories for Materials and Industries (Netherlands), 156
Lund Institute of Technology, Department of Structural Engineering (Sweden), 181

M

Maeda Construction Technical Research Institute (Japan), 145
Main Center for Building Information (Poland), 169
Manchester University, Institute of Science and Technology, Department of Building (United Kingdom), 207
Maryland, University of, 34, 48
Massachusetts Institute of Technology, 34–35, 49
Materials and Components Development and Testing Association (United Kingdom), 201
Materials Testing Institute, North-Rhine-Westphalia (West Germany), 112
Mediterranean Center for Research and Applied Studies in Industry and Construction (France), 123–124
Metal Building Manufacturers Association, 12–13
Michigan, University of, 35–36
Military Engineering Laboratory (Spain), 180
Ministry of Housing and Urbanism (Chile), 97–98
Ministry of Works and Development Central Laboratories (New Zealand), 160
Minnesota, University of, 36, 50
Montreal, University of, Faculty of Development (Canada), 95

N

Naples, University of, Department of Graphic Representation and Design Implementation (Italy), 141
National Association of Home Builders, 25
National Association of Home Builders Research Foundation, 13
National Association of Italian Engineers and Architects, 139
National Board of Physical Planning and Building (Sweden), 181–182
National Building Research Institute (South Africa), 177
National Buildings Organization (India), 133
National Cement Research Centre (Belgium), 85
National Center for Appropriate Technology, 72
National Center for Construction Laboratories (Iraq), 135
National Commission for the Civil Construction Industry (Brazil), 90
National Concrete Masonry Association, 13–14
National Construction Council (Tanzania), 191
National Glass Institute (Belgium), 86
National Gypsum, 23
National Housing Association (Netherlands), 156
National Housing and Building Research Unit (Tanzania), 191
National Institute of Building Sciences, 73
National Institute of Construction Management and Research (India), 134
National Institute of Housing (Venezuela), 210
National Institute for Physical Planning and Construction Research (Ireland), 136
National Institute of Technology, Department of Building and Construction (Norway), 163–164
National Institute for Transport and Road Research (South Africa), 177
National Laboratory of Civil Engineering (Portugal), 172
National Research Council Institute for Housing and Social Infrastructures (Italy), 139–140
National Research Institute for Construction and Construction Economics (Romania), 174
National Swedish Road and Traffic Research Institute, 182
National Technical University of Athens, Department of Theoretical and Applied Mechanics (Greece), 128–129
Naval Civil Engineering Laboratory, 63–64
Naval Explosive Ordnance Disposal Technology Center, 67
Naval Facilities Engineering Command, 65
NBD Product-Information Systems BV (Netherlands), 157
Netherlands Waterworks Testing and Research Institute, The, 159
New Brunswick University, Department of Civil Engineering (Canada), 96
Newcastle upon Tyne, School of Architecture, Building Science Section (United Kingdom), 207
Newcastle upon Tyne University, Department of Civil Engineering (United Kingdom), 208
New Mexico, University of, 50

New South Wales, University of (Australia), 80
New York, State University of, 36, 51
New Zealand Concrete Research Association, 160–161
New Zealand Heavy Engineering Research Association, 161
Nippon Telegraph and Telephone Corporation, Building Engineering Department (Japan), 145
North American Philips Lighting Corporation, 24–25
Northwestern University, 51–52
Norwegian Building Center, 164
Norwegian Building Research Institute, 164
Norwegian Geotechnical Institute, 165
Norwegian Institute of Wood Technology, 165

O

Ohbayashi-Gumi, Ltd. (Japan), 146
Ohio State University, 37
Ontario Buildings Branch (Canada), 94
Ontario Research Foundation (Canada), 94
Oregon State University, 52
Oregon, University of, 37
Otis Elevator, 28
Ove Arup and Partners (United Kingdom), 201
Owens-Corning Fiberglas, 24
Oxford Polytechnic, Department of Civil Engineering, Building and Cartography (United Kingdom), 202

P

Pennsylvania State University, 52–53
Philips International, BV, Building Design and Engineering Division (Netherlands), 157
Philippine Council for Industry and Energy Research and Development, 168
Philippines, University of The, Building Research Service, National Engineering Center, 167
Physical-Technical Institute for Research on Heat and Sound Technology (Austria), 83
Planning Committee for Building Research of The Netherlands, Organization for Applied Scientific Research, 157–158
Plymouth Polytechnic, Department of Civil Engineering (United Kingdom), 202
Polytechnic of the South Bank, Faculty of the Built Environment (United Kingdom), 202
Portland Cement Institute (South Africa), 178
Portsmouth Polytechnic, Department of Civil Engineering (United Kingdom), 203
PPG Industries, 25
Princeton University, 37–38
Property Services Agency (United Kingdom), 209–210
Public Works Canada, 94
Public Works Research Laboratory (France), 125
Purdue University, 53–54

Q

Qualitel (France), 126
Quality Declarations Organizations for Building Materials and Components (Komo), (Netherlands), 155–156
Queensland Institute of Technology, Faculty of the Built Environment, Department of Architecture and Industrial Design, 79

R

Redland Technology Limited (United Kingdom), 203
Rensselaer Polytechnic Institute, 38
Research Association for Building and Housing (West Germany), 110
Research Association for Underground Transportation Facilities, Inc. (West Germany), 112–113
Research Center for Civil Engineering (Czechoslovakia), 101–102
Research and Design Center for Industrial Building (Poland), 170
Research and Design Institute for Physical Planning, Housing, and Municipal Engineering (Romania), 175
Research and Experimentation Center for the Ceramic Industry (Italy), 138–139
Research Institute for Building and Architecture (Czechoslovakia), 101
Research Institute for Building Physics (USSR), 193–194
Research Institute of the Cement Industry, Association of the German (West) Cement Works, 110
Research Institute for Concrete and Reinforced Concrete (USSR), 193
Research Institute for Environmental Development (Poland), 170
Research Institute for Light Metals (Italy), 138
Research Institute of Mineral Building Materials (Poland), 170
Research Society for Housing, Building, and Planning, 81
Research Triangle Institute, 73
Residential Conservation Service, 68
RIW Institute for Housing Research (Netherlands), 158
Road and Bridge Research Institute (Poland), 171
Road Research Centre (Spain), 180
Road Research Unit, National Roads Board (New Zealand), 161
Rome University, Department of Industrial Design and Building Production (Italy), 141
Royal Danish Academy of Art, School of Architecture, 106
Royal Danish Academy of Fine Arts, School of Architecture, Institute of Building Science, Department of Building Science, 106
Royal Institute of British Architects, 203

Royal Institute of Technology, Division of Building Technology (Sweden), 183

S

Salford University, Department of Civil Engineering (United Kingdom), 208
San Carlos University, Engineering Research Center (Guatemala), 129
Sato Kogyo Co, Ltd. (Japan), 146
Schlage Lock Co., 25
Scientific and Technical Research Center (France), 124
Scientific and Technical Research Council of Turkey, Building Research Institute, 192
Shanghai Research Institute of Building Sciences (China), 99
Sheffield University, Department of Civil and Structural Engineering (United Kingdom), 208
Shimizu Construction Co., Ltd., Institute of Technology (Japan), 146–147
Singapore National University, Faculty of Architecture and Building, 176
Sintef Department of Architecture and Building Technology (Norway), 165
Sintef Norwegian Fire Research Laboratory, 166
Siporex Central Laboratory (Sweden), 183–184
Slovak Academy of Sciences, Institute for Construction and Architecture (Czechoslovakia), 102
Smart House Consortium, 25
Society of Heating, Air-Conditioning, and Sanitary Engineers of Japan, 147
Society for the Study of Precast Concrete Construction (Netherlands), 158
Sofus-Byg–Cooperation on Research, Development, and Technological Services in the Building Sector (Denmark), 106–107
Solar Energy Research Institute, 69–70
South of Portugal's Building and Public Works Association, 172
Southwest Research Institute, 73–74
SRI International, 74
Stanford University, 54–55
State University of Ghent, Faculty of Engineering, and Laboratory for Fuel Technology and Heat Transfer (Belgium), 87–88
State University of Liege, Research Centers for Architecture and Town Planning, Civil Engineering, Building Physics Laboratory (Belgium), 88–89
Steel Construction Institute (United Kingdom), 204
Steven Winter Associates, Inc., 26
Stevens Institute of Technology, 55
Strathclyde University, Department of Architecture and Building Sciences (United Kingdom), 209
Structural Engineering Research Center, Madras (India), 134
Stuttgart University, Geotechnical Institute for Underground Building, Soil Mechanics, Rock Mechanics, and Tunnel Construction (West Germany), 119
Stuttgart University, Institute for Concrete Structures (West Germany), 118–119
Stuttgart University, Institute of Construction Management, Institute of Structural Analysis (West Germany), 119
Sydney, University of, Department of Architectural Science (Australia), 80–81
Sylvania Products Corporation, 26
Syndicates of Reinforced Concretes and Industrialized Techniques (France), 126
Sweden Corrosion Institute, 184–185
Swedish Building Centre, 184
Swedish Cement and Concrete Institute, 184
Swedish Council for Building Research, 185
Swedish Environmental Research Institute, 185
Swedish Geotechnical Institute, 186
Swedish Institute for Steel Construction, 186
Swedish Institute of Building Documentation, Byggdok, 186
Swedish National Testing Institute, 183
Swiss Building Documentation Center, 189–190
Swiss Federal Institute of Technology, Applied Statics and Steel Structures; Institute of Structural Engineering; Institute for Building Materials, Material Chemistry, and Corrosion; Institute of Foundation Engineering and Soil Mechanics; Institute for Building Research, 189
Swiss Federal Institute of Technology, Department of Civil Engineering, 187
Swiss Federal Institute of Technology, Laboratory for Building Materials Science, 188
Swiss Federal Institute of Technology, Research Institute for the Built Environment, 187–188
Swiss Federal Laboratories for Materials Testing and Research, 188–189
Swiss Institute of Steel Construction, 190

T

Taft Laboratory, 70
Taisei Corporation, Engineering and Construction (Japan), 147–148
Takenaka Komuten Company, Ltd., Technical Research Laboratory (Japan), 148
Tamper University of Technology, Department of Civil Engineering (Finland), 122
Technical Association of the Cement Industry (Portugal), 172–173
Technical Center for Air Handling and Heating Industries (France), 125
Technical Control Bureau for Construction Safety (Belgium), 84
Technical Institute of Materials and Constructions (Spain), 179–180
Technical Research and Advisory Institute of the Swiss Cement Industry, 190

Technical University of Budapest, Faculty of Civil Engineering, Faculty of Architecture (Hungary), 131
Technical University of Denmark, 107
Technical University of Dresden, Sections of Civil Engineering, Architecture, Energy Transformation (East Germany), 127
Technical University of Graz—Institute for Soil Mechanics, Rock Mechanics, and Foundation Engineering (Austria), 83
Technical University of Vienna, Institute for Building Construction (Vienna), 83–84
Technological Centre of Minas Gerais (Brazil), 90–91
Tennessee Energy Conservation in Housing, 68
Tennessee, University of, 68
Texas, The University of, at Austin, 38, 55–57
3M Companies, 27
Tianjin Fire Research Institute (China), 99
Timber Center for Research, Information, and Education (Netherlands), 152–153
Timber Research and Development Association (United Kingdom), 204
Toda Construction Co., Ltd., Institute of Technology (Japan), 148
Transport and Road Research Laboratory (United Kingdom), 204–205
Transportation Research Institute (Israel), 137

U

Ulster, New University of (United Kingdom), 201
Underwriters' Laboratories of Canada, 95
United Architects of The Philippines, 168
United Nations, 133
United Technologies, 27–28

University College London, Bartlett School of Architecture and Planning (United Kingdom), 205
University College London, Department of Civil Engineering (United Kingdom), 205
USG Corporation, 27

V

Venezuela, Central University of, Institute of Materials and Structural Models, 210
Victoria University of Wellington, School of Architecture, 162
Virginia Polytechnic Institute and State University, 39

W

Wales, University of, Institute of Science and Technology, Welsh School of Architecture, Research, and Development (United Kingdom), 209
Washington, University of, 39, 57
Water Engineering Research Laboratory, 70
Waterloo University, Waterloo Construction Council (Canada), 96
Waterways Experiment Station (U.S. Army), 66
Western Ontario, University of, Boundary Layer Wind Tunnel Laboratory (Canada), 97
W.R. Grace & Co., 29
Wisconsin-Madison, University of, 58
Wisconsin-Milwaukee, University of, 40
Wood and Furniture Technical Center (France), 124
Wood Technology Institute (Poland), 171

Y

Yugoslav Building Center, 212

Index of Subjects

A

Accidents and accident prevention, 95
  see also Emergency systems and services
Acid rain, 58–59
Acoustics, 14, 15, 23, 24, 27, 29, 33, 52, 53, 56, 65, 81, 83, 84, 87, 93, 103, 111, 117, 124, 137, 140, 142, 146–148, 162–164, 169, 172, 173, 176, 177, 194, 197, 199, 201, 206, 207, 209
  glass, 20
  insulation, 83, 126
Additives, 29
Adhesives, 67, 71, 82, 108, 160, 165, 171
Administration, *see* Management and management services
Aerodynamics, 160
  see also Wind and wind tunnels
Aerospace industry, 74
Aesthetics, 19
Aged persons, 30
Agricultural structures, 100, 114, 127, 173
Air conditioning, 7, 20, 21, 23, 46, 52, 98, 111, 116, 125, 145, 198
Airports, 187
Air quality and pollution, 10, 185
  acid rain, 58–59
  asbestos, 70
  building materials, effects, 59, 108, 129
  indoor, 7, 13, 20, 46, 52, 59, 61, 68, 69, 71–72, 111, 147, 173, 197, 199
  see also Ventilation
Alloys, 138
Aluminum, 138
Architecture, 4, 18–19, 79, 80–81, 86–89, 101, 102, 106, 127, 131, 139, 140–141, 147, 149, 151, 153, 155, 157, 162, 165, 168, 174–176, 192, 193, 199, 200, 203, 205–207, 209
Artificial intelligence, 12, 15, 28, 33, 49, 52

  see also Expert systems
Asbestos, 70
Asphalt, 7–8, 211
Automation, *see computer terms*; Environmental control systems; Industrial process control; Robotics
Automobiles, 177
  glass, 19, 86
Awards, 9, 143

B

Ballistics, 24
Basements, 54
Behavioral sciences, 30, 35, 36, 36, 40, 48, 64, 181, 188, 189
  energy conservation and, 40
  labor and employment, 49, 90, 176, 181, 182
  lighting, 12
  residences, 31, 33, 170, 191
  see also Aged persons; Handicapped persons; Management and management services; Social factors
Biodegradation, 67
Biotechnology, 147, 160
Blast loads, 74
Bonding, *see* Adhesives
Bricks, 8–9, 78, 93, 104–105, 121, 143
  reinforced, 199, 202
Bridges, 5, 53, 54, 89, 94, 101, 114, 115, 116, 122, 125, 141, 161, 171, 186, 211
Bronze, 59
Building codes, 26, 38, 39, 93, 94, 98, 107, 113–114, 121, 139, 150, 164, 177, 180–182, 197
  concrete, 118, 123, 158
  fire, 99
  seismic safety, 42

steel structures, 186
see also Inspection services
Building energy design, *see* Energy consumption and conservation; Heating systems; Solar energy
Building physics, 83, 84, 86, 87–89, 91, 98, 99, 102, 103, 107, 108, 112, 113, 117, 127, 130, 157, 169, 174, 183, 192–194, 201

C

CAD/CAM (Computer-aided design/manufacturing), 14, 16, 18, 23, 26, 30, 33, 35, 36, 37, 39, 40, 48–49, 49, 52, 53, 55, 61, 64, 65, 137, 147, 150, 165, 170, 193, 199, 200, 209
  architecture, 153, 162
  cement and concrete, 103
  civil engineering, 114, 181, 195, 203, 208
  explosives, terrorists, 17
  home building, 31
  plywood building design, 6
  steel, structural, 5, 51, 55, 92
  structural, 14
  wood, 108
  see also Robotics
Calibration, 90, 105, 107, 114, 125, 182, 183
Canals, 87, 109
Cartography, 202
  photogrammetry, 45, 122
Case studies, 60
Ceilings, 14
Cement, 9, 44, 45, 55, 62, 78, 81–82, 85, 100, 110, 129, 137, 138, 154, 170, 173, 178, 179, 190, 191, 197–198, 211
  additives, 29, 179
  portland, 78
  structural, 29, 56, 66, 102–103
Ceramics, 16, 17, 44, 129, 138, 196, 203, 206
Chemicals and chemical analyses, 8, 15, 22, 23, 24, 28, 60, 62, 63, 90, 105, 109, 111, 131
  bricks, 105
  cement and concrete, 110, 117
  electrochemistry, 85
  glass, 86
  indoor environments, 72
  metals, 103
  wood, 67, 82, 171
  see also Coatings; Corrosion; Paint and varnish
Children, 36
City planning, *see* Town planning; Urban planning
Civil engineering, 6–7, 18, 34, 39, 44, 45, 47, 49, 53, 57, 63, 80, 84, 85, 87, 88–91, 96–97, 101, 109, 113, 116, 117, 120–122, 124–127, 136, 137, 140, 144–147, 149, 154, 155, 157, 164, 168, 171–174, 178, 182, 189, 192–195, 197–200, 201–206, 208, 211
  education, 96, 121, 122, 126, 128, 162, 178, 179, 181, 187, 190
  see also Bridges; Dams; Roads; Rock mechanics; Soil mechanics; Traffic engineering; Transportation
Climate, 7, 13, 37, 44, 56, 137, 142, 194
Coal, 156
Coastal zones, 107, 118, 153, 202
Coatings, 22, 25, 44, 60, 62, 63, 71, 72, 86, 185
  glass, 19, 20
  roofs, 126
  wood, 42, 51, 52, 63
  see also Paint and varnish
Codes, *see* Building codes
Cold environments, 41, 64, 120
Communications, 21, 22, 35, 157, 197
  see also Telecommunications
Composite materials, 15, 42, 46, 58, 60, 63, 67, 83, 116, 187
  see also Fiber composites
Computer control systems, 11–12, 147, 169, 174, 201
  fire and security, 21, 22
  industrial processes, 3, 21, 94, 95, 108
  see also Artificial intelligence
Computer programming, 15, 45, 98, 115, 119, 178, 194
  building codes, 94
  civil engineering, 181
  concrete, 103
  energy analysis, 3
  lighting, 104
  management, 106
  steel, 92
  see also Artificial intelligence; CAD/CAM; Data bases; Expert systems
Computer simulation, 48, 50, 56, 58, 63, 65, 207
  concrete, 134
  energy storage and use, 39, 68
  fire, 72
  historic structures, 38
  steel structures, 116
  see also CAD/CAM; Expert systems
Concrete, 42, 44, 45, 46, 47, 48, 51, 52, 53, 55, 56, 58, 60, 63, 65, 66, 78, 79, 80, 82, 85, 96, 102, 103, 110, 114, 116, 117, 120, 123, 128, 133–135, 137, 140, 146–148, 154, 158, 160–161, 164, 167, 172, 177–179, 183, 184, 188, 190, 193, 195, 197–198, 202, 203, 205–207, 211
  additives, 29, 179, 191
  fire and explosion effects, 72
  lightweight, 174
  plastic, 83
  prestressed, 77, 87, 115, 117, 118, 174, 179, 189
  reinforced, 85, 87, 89, 102, 113, 115, 117, 118, 126, 128, 133, 134, 174, 187, 189, 193
  roads, 162
  standards and testing, 4, 9, 13, 131
  structural, 29, 56, 66, 102–103, 109, 113, 114, 122, 126, 193, 199, 205, 210
Conservation, *see* Historical conservation; Rehabilitation and renovation

Construction equipment, 16, 49, 53, 84, 96, 97–99, 105, 124, 126, 143, 144, 145, 148, 169, 171, 175, 178
Consumers and consumption, 72, 104, 126, 188
  see also Marketing
Contracts, 4, 80, 107
Cooling systems, 23, 43, 58, 60, 63, 69
  see also Air conditioning; Heat pumps; Refrigeration
Cork, 171
Corrosion, 44, 63, 74, 85, 86, 104, 109, 156, 157, 183–186, 189, 193
Cost analyses, 26, 39, 46, 53, 56, 58, 61–62, 71, 91, 119, 131, 137, 150, 167, 210
  electricity, 10
  home building, 13, 133, 166, 191, 205, 210
  rehabilitation and maintenance, 32
Curtain walls, 22

D

Dams, 165
Data bases, 115, 130, 140, 178, 186, 190, 194, 210, 212
  standards, 3, 14
  underground transportation, 113
Daylight, 35, 37, 38, 39, 68, 106, 165
Demography, 90
Demolition, 16–17
Detection, see Security systems
Developing countries, 30, 97
Diesel fuel and engines, 16
Disabled, see Handicapped persons
Disasters, 30, 134, 144, 148
  see also Emergency systems and services
Documentation, see Information systems
Doors, 22

E

Earth movers, 16
Earthquake engineering, see Seismic safety
Economics, 48, 61, 65, 90, 91, 101, 107, 123, 127, 136–139, 152, 162, 164, 173, 174, 176–178, 186–189, 193, 194, 199, 201–203, 205, 207
  architectural, 19
  building, 98, 103, 110, 122, 124, 144
  civil engineering, 172, 179, 195
  fire, 104
  housing, 13, 91, 110, 144, 158, 182, 188, 191, 205, 210
  solar energy, 43
  wood, 171
  see also Cost analyses; Financial planning; Marketing; Productivity
Education, 4, 32, 48, 79, 80, 83, 85, 87–89, 92, 95, 98, 101, 103, 106–107, 109, 118, 127–129, 136, 137, 141, 147, 152, 159, 175, 176, 180, 188, 189, 199, 200, 205, 209, 212
  air handling, 125
  architecture, 106, 127, 139, 140–141, 153, 162, 168, 205–207
  brick masonry, 8
  cement and concrete, 77, 103, 117, 123, 133, 161, 178, 198
  ceramics, 138
  civil engineering, 96, 121, 122, 126, 128, 162, 178, 179, 181, 187, 203
  coatings, 86
  environmental engineering, 185, 195, 209
  fire safety, 104, 117
  information systems, 194
  management, 119, 134
  plastics, 82
  soil and rock mechanics, 120
  steel, 77, 92, 116, 161, 190
  wood, 108
Elderly persons, see Aged persons
Electrical engineering, 18, 24, 157, 198
  cost-effectiveness, 10
  diesel-electric sets, 16
  power control and generation, 23, 74
  solid fuel cells, 28
  wiring, 13, 21, 126
  see also Solar energy
Electrochemistry, 85
Electronics, 28, 133, 147, 163, 183
  see also Communications, Telecommunications
Elevators, 131
Emergency systems and services, 12, 30, 37, 94, 95, 144
  see also Fire safety; Security systems
Employment and labor, see Labor and employment
Energy consumption and conservation, 3, 10, 33, 35, 36, 38, 39–40, 46, 50, 52, 53, 60, 63, 64, 67, 69, 72, 74, 80, 81, 85, 88, 116, 117, 119, 125, 129, 140, 142–144, 146, 151, 157, 159, 162, 164, 165, 167, 168, 176, 177, 179, 182, 183, 185, 186, 189, 191, 192, 198–201, 203, 207, 209
  coatings, 25
  fuel, 189
  glass and, 19, 71
  lighting, 24, 26, 33, 36
  modeling, 26
  retrofitting, 3, 21, 68
  storage, 39, 43, 52
  transportation, 137
  wind, 97
  see also Electrical engineering; Solar energy
Environmental control systems, 37, 39, 45, 46, 47, 51, 52, 55, 91, 92, 111, 144, 145, 201
  see also Air conditioning; Air quality and pollution; Cooling systems; Heating systems; Insulation; Ventilation
Environmental sciences, 35, 40, 48, 50, 64, 79, 85, 95, 96, 101, 107, 117, 123, 125, 139, 141, 145–148, 151, 152, 154, 156, 157, 161, 165, 168, 169, 177, 183, 185, 186, 195, 199, 200, 203, 206–209

cement and concrete, 110
electrical generators, 10
housing and, 170
transportation, 109, 112
see also Air quality and pollution; Sanitation; Waste management
Equipment, *see* Construction equipment
Ergonomics, *see* Behavioral sciences
Erosion, 72
Error analyses, 48
Expert systems, 30, 32, 46, 49, 81, 137, 181
Explosives and explosions, 19, 50, 66, 67, 71–72, 74, 99, 157
see also Demolition

F

Fiber composites, 14, 27, 44, 102
   ceramic-carbon, 17
   fiberglass, 24
   plastics, 39, 83
Financial planning, 49, 108, 156
   funding promotion, 4
   office buildings, 9
Fire safety, 30, 50, 63, 71–72, 73, 93, 94, 95, 98, 111, 117, 123, 124, 126, 131, 132, 144, 147, 148, 157, 159, 166, 172–174, 177, 183, 192, 197, 201, 204, 211
   control systems, 18, 21, 22, 55, 99, 104, 105, 113, 193, 194
   retardants/resistants, 14, 22, 23, 29, 193
   steel, 5, 186
Flooding, 37
Flooring, 42, 45, 47, 55, 57, 196
   fire retardants, 14
Foam materials, 15, 22
Forestry, 39, 123, 160, 167
   see also Wood and wood products
Foundations, 40–41, 50, 51, 56, 63, 64, 83, 98, 107, 108, 114, 116, 119, 120, 123, 127, 128, 135, 136, 141, 144–148, 154, 160, 162, 165, 167, 172, 182, 186, 189, 198, 206, 211
Fuel, 189
Furnaces, 71, 74
Furniture, 14, 59, 124, 141

G

Gas, 3, 11
Geography, 50
Geotechnical engineering, 47, 50, 51, 54, 56, 57, 66, 74, 96, 122, 125, 132, 133, 141, 145, 147, 151, 153–154, 165, 167, 169, 171, 172, 174, 177, 179, 186, 195, 197, 200, 203, 205, 206, 208
   see also Foundations; Rock mechanics; Seismic safety; Soil mechanics
Girders, 3
Glass, 22, 25, 69, 71, 86, 129, 164
   architectural, 19, 20, 25
   automotive, 19, 86
   ceramic, 17
   see also Windows
Glue, *see* Adhesives
Granite, 72
Gypsum, 23, 27, 29, 44, 156, 167, 170, 191

H

Handicapped persons, 30, 36, 44, 71
Harbors, 118
Hazardous substances
   asbestos, 70
   waste, risk analyses, 3
   see also Air quality and pollution, indoor
Health effects, 59, 143, 206
   see also Behavioral sciences
Heat pumps, 60, 68–69, 201
Heating systems, 7, 20, 21, 23, 43, 46, 52, 53, 58, 60, 71, 83, 107, 116, 123, 125, 127, 169, 198
High-rise buildings, 34, 48
Highways, *see* Roads
Historical conservation, 31, 32, 33, 36, 38, 38–39, 59, 86, 106, 149, 173
Home construction and repair, 31, 37, 53, 72, 81, 91, 97–98, 100, 108, 111, 126, 127, 130, 133, 134, 152, 158, 163, 166, 155–157, 170, 176, 177, 185, 188, 189, 207–209
   cement, 77
   economics, 13, 91, 110, 144, 158, 182, 188, 191, 205, 210
   energy, 43, 60
   prefabricated, 34
   quality control, 91, 133, 163, 182, 210
Human factors, *see* Behavioral sciences
Humidity, 7, 13, 43
Hydraulics, 45, 47, 55, 57, 89, 101, 118, 122, 125, 145, 147, 153, 154, 160, 172, 179, 199, 202, 203, 205, 208
   binders, 190
   river flows, 87
Hydrodynamics, 147
Hydrology, 96, 107, 202, 206

I

Illumination, *see* Lighting
Industrial buildings, 114, 127, 141, 161, 169, 173, 178
Industrial process control, 3, 21, 22, 94, 95, 108, 171, 175, 183
   acoustics, 111
   cement, 110
Information systems, 3, 14, 84, 90, 92–94, 99, 103, 111–112, 115, 121, 125–128, 130, 131, 132, 137–139, 143, 147, 150–153, 157, 159, 164, 169, 174–178, 181, 184–186, 188, 190, 192, 194, 196–198, 209, 210, 212
   architecture, 168, 203
   cement, 173
   civil engineering, 125, 182, 195, 211
   housing, 135, 182, 191, 209

wood, 124, 153
see also Data bases; Statistical programs and activities
Infrastructure, *see* Civil engineering
Inks, 86
Inspection services, 4, 46, 100, 125, 131
  brick, 78
  fire safety, 73, 104, 131
  waterworks, 159
Insulation, 13, 14, 22, 24, 27, 53, 62, 71, 83, 196
  acoustic, 83
  glass, 20
  residential superinsulation, 72
  thermal, 83, 117, 126, 140, 142
  wood, 171
Intelligent buildings, 12

K

Knowledge-based systems, *see* Artificial intelligence; Expert systems

L

Labor and employment, 49, 90, 176, 181, 182
Lamination, *see* Coatings
Land management, 94, 123, 137, 182
  see also terms under Planning
Landscaping, 95
Laser technology, 133
Law, 107
  see also Building codes; Regulations; Standards
Licenses and permits, 101
Lifts, *see* Elevators
Lighting, 12, 14, 20, 21–22, 37, 39, 40, 41, 52, 62, 80, 81, 104, 162, 164, 172, 173, 176, 177, 194, 197, 198, 206, 207, 209
  daylight, 35, 37, 38, 39–40, 68, 106, 165
  energy efficiency, 24, 26, 33, 36
Lime, 167, 170
Loads (forces), 50
  glass and ceramics, impact, 17, 25, 164
  pilings, 56
  see also Seismic safety; Structural engineering and loads

M

Maintenance, 32, 49, 127, 139, 145, 160
  housing, 156, 157, 170
  roads, 161, 180, 202
Management and management services, 49, 64, 80, 81, 92, 94, 96, 106, 107, 134, 137, 144, 147, 151, 152, 155–157, 172, 176, 185, 188, 191–193, 195, 199, 202, 208, 209
  education, 119
  laboratory, 124
  office buildings, 9
  roads, 161, 180
  telecommunications, 145
  water, 125

  see also terms listed under Planning
Manufacturing processes, *see* Industrial process control
Marine engineering, 66, 87, 97, 107, 118, 125, 144, 147, 148, 155, 165, 198, 199, 202, 205
Marketing, 90, 139, 174, 177
  cement and concrete, 78, 154
  electricity use, 10
  housing, 91, 188
  intelligent buildings, 11–12
  plywood, 6
  steel, 77, 92
Masonry, 42, 55–57, 87, 93, 110, 114, 117, 120, 128, 138, 141, 155, 196, 198, 206
  see also Bricks; Cement; Concrete
Mass transfer, 69
Materials, *see specific materials*
Mathematical models, 62, 66, 89, 116
  soil and rock mechanics, 120
Measurement, *see* Calibration; Standards
Metals, 12, 16, 29, 42, 51, 58, 63, 102, 103, 109, 113, 155, 157, 189
  light, 116, 138
  sheet, 23
  see also Corrosion; Welding; *specific metals*
Meteorology, *see* Climate
Military facilities, 63–65, 180
Mines and mining, 19, 156, 169, 189
Mirrors, 86
Mobile homes, 31, 69
Models, 106, 210
  aerodynamic, 160
  fire, 99
  hydraulic, 87, 118, 153, 160
  see also Computer simulation; Mathematical models; Theoretical studies
Modular components, *see* Prefabricated structures
Moisture control, *see* Waterproofing
Mortar, 118, 173, 190

N

Natural gas, 3
Noise, *see* Acoustics
Nondestructive testing, 109, 124, 189, 206, 207, 211
North Sea, 87
Nuclear facilities and weapons, 17, 66
Numerical analyses, *see* Mathematical models

O

Occupational health and safety, 181, 182
Ocean engineering, *see* Marine engineering
Office buildings, management, 9
Offshore structures, *see* Marine engineering
Optics, 17, 28, 63
Organic materials, 109
Organizational services, *see* Management and management services

P

Paint and varnish, 58–59, 62, 63, 82, 86, 98, 114, 125, 157, 164, 177, 185
Paper and paper products, 23, 82
Particle board, 51
Pavements, 7, 50, 63, 64, 78, 180, 187, 203, 211
Pedestrians, 112
Petroleum, 8
Photogrammetry, 45, 122
Photography, 32, 47
Photovoltaics, 19
  see also Solar energy
Pipes, 55, 60, 145
  cement and concrete, 9
  gas, 3
  steel, 116
  water, 7–8
Planning, *see* Land management; Management and management services; Regional planning; Space planning; Town planning; Urban planning
Plaster, 27, 44
Plastics, 6, 15, 23, 29, 39, 53, 82–83, 113, 157, 172, 189
Plumbing, 7–8, 55, 62, 126, 147
Plywood, 6, 51
Policy planning, *see specific subject area*
Pollution, *see* Air quality and pollution; Environmental sciences
Polymers, 24, 60, 74, 114, 203
  resins, 24, 60, 171
Population studies, *see* Demography
Prefabricated structures, 34, 48, 116, 151, 181, 212
  cement and concrete, 77, 175
Prizes, *see* Awards
Productivity, 95, 132, 173, 207, 210
Psychology, *see* Behavioral sciences
Public works, *see* Civil engineering

Q

Quality control, 34, 62, 84, 90–91, 94, 95, 98, 100, 101, 105, 112, 118, 120, 126, 129, 130–131, 135–137, 140, 141, 144, 146, 155–156, 158, 176, 179–180, 189, 191–193, 196, 198, 201, 202, 212

  cement and concrete, 102, 113, 123, 146, 178
  civil engineering, 6, 34, 84, 113, 114, 121, 181, 200, 210, 211
  computer software, 115
  housing, 91, 133, 163, 182, 210
  masonry, 93, 120, 206
  wood, 124, 160, 171, 196, 204
  see also Building codes; Calibration; Inspection services; Regulations; Standards
Quarries, 169

R

Radiation, 147
Radon, 59
Railroads, 112, 154
Real-time computer system, 6
Recycling, 124, 143, 146, 155, 156
Refractory materials, 17, 44, 102, 196
Refrigeration, 7, 170
Regional planning, 86, 103, 108, 110, 111, 133, 205
  see also Land management; Rural planning; Town planning; Urban planning
Regulations, 94, 97, 114, 121, 127, 131, 139, 141, 156, 164, 178, 189
  intelligent buildings, 12
  steel structures, 186
  see also Building codes; Standards
Rehabilitation and renovation, 9, 23, 32, 55, 81, 87, 100, 106, 123, 127, 139, 141, 145, 152, 155, 165, 189, 200
  concrete structures, 66
  roads, 109
  see also Historical conservation; Home construction and repair
Repair, *see* Home construction and repair; Rehabilitation and renovation
Residential construction, *see* Home construction
Resins, 24, 60, 171
Retrofitting
  energy, 3, 21, 68
Risk analyses, 26, 33, 54, 58, 63, 95, 157, 181
  hazardous wastes, 3
Rivers, 87, 125, 153
Roads, 78, 87, 94, 101, 109, 114, 122, 123, 125, 128, 131, 132, 135–137, 150, 154, 171, 177, 179, 204, 205, 211
  cement and concrete, 77, 187
  lighting and marking, 104
  maintenance, 161, 180, 202
  pavements, 7, 50, 63, 64, 78, 180, 187, 203, 211
  urban, 112
Robotics, 48–49, 124, 130, 137
Rock mechanics, 66, 83, 89, 101, 107, 117, 119, 120, 125, 145, 147, 148, 160, 165, 169
Roofing, 8, 22, 29, 42, 45, 51, 62, 63, 68, 71, 126, 143, 196
Rubber, 29, 157
Rural planning, 98, 102, 106, 114, 132, 182

S

Saline systems, 117
Sanitation, 47, 96, 122, 129, 196
Security systems, 18, 21, 22, 25, 28, 63, 131
Seismic safety, 3, 17, 37, 41, 48, 54–57, 61, 74, 98, 115, 116, 128, 141, 144, 146, 147, 159, 162, 173, 174, 189, 192, 194, 210
  cement and concrete, 9, 45, 55, 56, 115, 199
  codes, 42
  developing countries, 30–31
  steel, structural, 5, 12, 45, 55, 56, 116, 199

Sensing technology, 21, 28
  see also Fire safety; Security systems
Silicates, 17, 129
Social factors, 38, 90, 102, 127, 136, 140, 145, 187–188, 191
  see also Behavioral sciences; Management and management services
Software, *see* Computer programming
Soil mechanics, 50, 54, 56, 66, 83, 89, 101, 107, 116, 119, 120, 122–128, 131, 135, 137, 144–148, 154, 160, 180, 186, 187, 189, 191, 199, 199, 200, 202, 206, 211
  Underground construction and structures
Solar energy, 37, 43, 45, 58, 61, 67, 69, 117, 165
  passive, 3, 39–40, 55, 173, 183
  see also Photovoltaics
Solid mechanics, 96
Space planning, 38, 155, 209
Standards, 8, 63, 65, 79, 90, 93, 95, 98, 107, 109, 113–114, 121, 127, 129, 131, 136, 138, 140, 141, 143, 150, 151, 163, 168, 172, 176, 189, 207
  cement and concrete, 4, 78, 122, 123
  data bases, 3, 14
  fire safety, 73, 99
  lighting, 25
  masonry, 110
  steel structures, 186, 190
  thermal materials, 142
  wood, 5, 124
  see also Building codes; Calibration
Statistical programs and activities
  civil engineering, 179
  demography, 90
  fire, 99
  office building management, 9
  transportation, 177
Steel, 27, 42, 53, 55, 77, 96, 113, 114, 147, 204
  CAD/CAM, 5, 51, 55, 92
  concrete reinforcement by, 85, 87, 116, 187
  loads, 12–13
  structural, 5, 42, 45, 46, 47, 51, 58, 92, 109, 113, 115–116, 161, 174, 186, 187, 189, 190, 199, 208
  wallboard and, 23
Stone, 72, 135, 143, 156
Storage
  energy, 39, 43, 52, 69
  gas, 11
  water, 173
Structural engineering and loads, 24, 42, 44, 45, 47, 48, 50–55, 57, 58, 61, 62, 63, 74, 87, 88, 91, 93, 96, 98, 99, 100, 103, 105, 107, 109, 111, 115, 117, 119, 122–124, 127–131, 136, 137, 141, 144, 145, 147, 149, 151, 160, 162, 169, 171, 174, 177, 179, 181, 198, 201–203, 205, 206, 208, 210, 211
  CAD, 114
  cement and concrete, 29, 56, 66, 102–103, 118, 122, 126, 189, 193, 199, 205
  fatigue, 46, 47, 57
  fracture, 47, 102, 155, 188
  steel, 5, 42, 45–47, 51, 58, 92, 109, 113, 115–116, 161, 174, 186, 187, 189, 190, 199, 208
  transportation, 161
  wood, 51, 67, 82, 108, 153, 165, 189, 204, 206
  see also Seismic safety; Wind and wind tunnels
Structural materials, 9, 41, 42, 46, 87–89, 92, 100, 103, 111, 146, 183, 191
  brick masonry, 8–9, 206
  cement and concrete, 29, 56, 66, 102–103, 118, 122, 126, 189, 193, 199, 205
  demolition, 16–17
  plastics, 6
  steel, 5, 42, 45–47, 51, 58, 92, 109, 113, 115–116, 161, 174, 186, 187, 189, 190, 199, 208
  wood, 51, 67, 82, 108, 153, 165, 189, 204, 206
Surveying, 122, 208
  photogrammetry, 45, 122

T

Telecommunications, 35, 145
Terrorism, 17
Testing, *see* Quality control
Theoretical studies, 36, 40, 53, 86, 100, 127, 170
  fire, 99
Thermal properties and performance, 80, 81, 87, 88, 117, 127, 131, 137, 147, 142, 173, 176, 183, 194, 196, 197, 200, 201, 211
  cold environments, 41, 64, 120
  glass, 71, 86
  insulation, 83, 117, 126, 140, 142
  soil, frozen, 120
  see also Heat pumps; Heating systems; Refractory materials
Third World, *see* Developing countries
Timber, *see* Wood and wood products
Total building systems, 30
  steel, structural, 5
Towers, 116
Town planning, 88–89, 94, 95, 101, 108, 110, 111, 114, 123, 127, 139, 154, 175, 202
Traffic engineering, 78, 83, 109, 123, 128, 137, 180, 182, 195, 208
  urban, 112
Translation (languages), 130
Transportation, 40, 78, 96, 97, 122, 128, 132, 133, 172, 177, 180, 203, 204, 208, 211
  fire safety, 74
  gas, 11
  railroads, 112, 154
  see also Bridges; Roads
Tunnels, 56, 101, 112, 117, 119

U

Underground construction and structures, 50, 54, 56, 64–65, 108, 112, 113, 117, 120, 146, 147, 198
  see also Foundations; Soil mechanics; Tunnels

Underwater construction and structures, *see* Marine engineering
Urban planning, 35, 36, 40, 48, 50, 86, 95, 97, 98, 101, 102, 103, 106, 141, 152, 158, 167, 182, 185, 188, 205
   traffic, 112
Utilities, 112

V

Varnish, *see* Paint and varnish
Ventilation, 7, 20, 21, 23, 46, 52, 63, 99, 123, 125, 127, 140, 177, 197, 198
   ceilings, 14
   gas appliances, 72
Vibration, 53, 63, 93, 124, 148, 163
Videotapes, 32, 49
Vinyl, 23

W

Wallboard, 23, 27, 44
Walls, 196
   indoor, 42, 52, 54, 55, 57, 62
   retaining, 120
War, 66
Waste management, 114, 185
   hazardous, risk analyses, 3
   industrial, 146
   recycling, 124, 143, 146, 155, 156
   water, 164, 203
   wood, 124
Water resources and engineering, 7–8, 45, 96, 122, 125, 147, 148, 151, 159, 199
   control systems, 20
   dams, 165
   flow models, 87
   supply and drainage, 55, 57, 85, 164, 173, 182
   waste, 164, 203
   see also Coastal zones; Hydraulics; Hydrology; Marine engineering; Plumbing; Rivers
Water vapor, 87
Waterproofing, 8–9, 22, 55, 62, 117, 159, 183, 194, 196
Waterways, 87, 109
Wave processes, 97, 118
Weapons systems, 66
Weather, *see* Climate
Welding, 5, 23, 46, 47, 161, 211
Wind and wind tunnels, 30, 35, 37, 39, 42, 63, 97, 147, 160, 203, 206
Windows, 22, 41, 55, 67, 196
Wood and wood products, 27, 31, 45, 51–53, 57, 58, 82, 108, 124, 152–153, 155, 160, 162, 165, 167, 171, 172, 195, 196, 204
   coatings, 42, 51, 52, 63
   plywood, 6, 51
   structural, 51, 67, 82, 108, 153, 165, 189, 206
   wallboard, 23, 52
   see Forestry; Paper and paper products